T0305491

GENTRIFICATION TRENDS IN THE UNITED STATES

Gentrification Trends in the United States is the first book to quantify the changes that take place when a neighborhood's income level, educational attainment, or occupational makeup outpace the city as a whole – the much-debated yet poorly understood phenomenon of gentrification. Applying a novel method to four decades of U.S. Census data, this resource for students and scholars provides a quantitative basis for the nuanced demographic trends uncovered through ethnography and other forms of qualitative research. This analysis of a rich data source characterized by a broad regional and chronological scope provides new insight into larger questions about the nature and prevalence of gentrification across the United States.

- Has gentrification become more common over time?
- Which cities have experienced the most gentrification?
- Is gentrification widespread, or does it tend to be concentrated in a small number of cities?
- Has the nature of gentrification changed over time?

Ideal reading for courses in real estate, urban planning, urban economics, sociology, geography, econometrics, and GIS, this pathbreaking addition to the urban studies literature will enrich the perspective of any scholar of U.S. cities.

Richard W. Martin, PhD, is Associate Professor of Real Estate at the University of Georgia's Terry College of Business. He is the co-author, with Raphael Bostic, of the landmark paper "Black Homeowners as a Gentrifying Force."

GENTRIFICATION TRENDS IN THE UNITED STATES

Richard W. Martin

Routledge
Taylor & Francis Group

NEW YORK AND LONDON

Designed cover image: © Shutterstock

First published 2024
by Routledge
605 Third Avenue, New York, NY 10158

and by Routledge
4 Park Square, Milton Park, Abingdon, Oxon, OX14 4RN

Routledge is an imprint of the Taylor & Francis Group, an informa business

Library of Congress Cataloging-in-Publication Data
Names: Martin, Richard W., author.
Title: Gentrification trends in the United States / Richard W. Martin.
Description: First Edition. | New York : Routledge, 2024. | Includes
bibliographical references and index.
Identifiers: LCCN 2023009096 (print) | LCCN 2023009097 (ebook) | ISBN
9781032107042 (paperback) | ISBN 9781032108872 (hardback) | ISBN
9781003217459 (ebook)
Subjects: LCSH: Gentrification--United States. | City
planning--Environmental aspects--United States. | Equality--United
States. | Social sciences--Research--Methodology.
Classification: LCC HT175 .M37 2024 (print) | LCC HT175 (ebook) | DDC
307.3/4160973--dc23/eng/20230501
LC record available at https://lccn.loc.gov/2023009096
LC ebook record available at https://lccn.loc.gov/2023009097

ISBN: 9781032108872 (hbk)
ISBN: 9781032107042 (pbk)
ISBN: 9781003217459 (ebk)

DOI: 10.1201/9781003217459

Typeset in Times New Roman
by KnowledgeWorks Global Ltd.

CONTENTS

1

INTRODUCTION

Introduction

This book is primarily concerned with the quantitative measurement of gentrification activity in U.S. cities from 1970 to 2010. The main goal of the book is to provide an estimate of the amount of gentrification that occurred in U.S. cities during that period of time. The analysis contained in the book will provide a general sense of just how common gentrification has been in U.S. cities, as well as the extent to which gentrification has become more common over time. Additionally, the cities that have been most affected by gentrification in each decade, as well as over the entire 40-year period, will be identified.

The primary contribution of this book to the gentrification literature is to provide a historical, cross-sectional study of the amount of gentrification occurring in U.S. cities. Specifically, the sample used in this book includes the 100 largest cities according to their populations in 1970 and allows for analysis across four decades. Applying the same methodology across a large sample of cities and across multiple decades provides a consistent estimate of the amount of gentrification that was taking place over time in U.S. cities.

An additional contribution of this book is that it measures three different types of gentrification in each city and decade. Separate chapters are devoted to measuring income, educational, and occupational gentrification and, since each type of gentrification is measured using the same methodology, it is possible to identify things such as which types of gentrification were most common at each point in time, how the levels in each type of gentrification change over time, and which cities were most affected by the different types of gentrification.

DOI: 10.1201/9781003217459-1

Overview of the Book

Following this introductory chapter, Chapter 2 will provide a brief overview of the literature related to using quantitative analysis to identify gentrifying neighborhoods. While the amount of published research on gentrification is massive, studies that use quantitative methods to identify gentrifying neighborhoods are a relatively small portion of the overall literature. The focus of Chapter 2 is on summarizing the various approaches that have been used and evaluating the strengths and weaknesses of each approach.

Chapter 3 will analyze the trends in population, employment, income, education, and occupation in U.S. metropolitan areas from 1970 to 2010. The purpose of Chapter 3 is to establish the overall economic environment within which the gentrification activity that is identified in the subsequent chapters is occurring. The focus is on such things as changes in the central city share of metropolitan population and employment, how central city income levels have changed relative to the levels of their metropolitan areas, and how central city educational attainment and occupational mix compare with their metropolitan areas. The primary purpose of the analysis in Chapter 3 is to identify whether the overall metropolitan socioeconomic trends from 1970 to 2010 were such that an increase in gentrification would have been expected or, as is found to be the case, did the increase in gentrification activity chronicled in this book occur in spite of the prevailing trends.

Chapter 4 is the beginning of the four chapters that contain the primary analysis in the book. The chapter begins by detailing the methodology that is used to identify gentrifying neighborhoods. Once the methodology has been explained, the rest of the chapter involves using the methodology to measure the amount of income gentrification in U.S. cities from 1970 to 2010. Chapter 5, then, involves using the same methodology to measure the amount of educational gentrification over the same timeframe, while Chapter 6 uses the methodology to measure the amount of occupational gentrification.

The analysis in Chapters 4–6 consisted of considering only one dimension of gentrification at a time. Chapter 7 uses the analysis from the previous three chapters to determine how common the various combinations of gentrification were over time. With three ways for a neighborhood to gentrify (by income, education, or occupation), there are several different ways for a neighborhood to gentrify: along only one dimension (income, education, or occupation), along two dimensions (income-education, income-occupation, and education-income), and, finally, a neighborhood could gentrify along all three dimensions. Chapter 7 identifies which gentrification combinations were the most common and least common and how the relative frequencies of the various types of gentrification changed over time.

Finally, Chapter 8 concludes the book by summarizing the primary findings from the analysis that was conducted throughout the book. The chapter also identifies several important unanswered questions that can provide a fruitful avenue for future research pertaining to the quantitative analysis of gentrifying neighborhoods.

2

USING QUANTITATIVE METHODS TO IDENTIFY GENTRIFYING NEIGHBORHOODS

A Survey

British sociologist Ruth Glass is typically credited with the first use of the term *gentrification* in its modern sense (Glass, 1964). In the time that has passed since her original use of the term, a massive amount of gentrification research has been conducted. While there have been very fruitful, and sometimes heated, debates over fundamental issues such as the primary causes of gentrification or the level of displacement that is associated with gentrification, this chapter will focus primarily on the issue of how gentrification has been measured in the relatively small portion of the literature devoted to quantitative analysis of gentrification. For readers interested in a more complete treatment of the entire gentrification literature, Lees et al. (2008) provide a very comprehensive treatment of the early development of the literature. Also, the articles collected in Brown-Saracino (2010) provide a good introduction to many of the important debates in the gentrification literature.

What Is Gentrification?

Before one can measure gentrification, it is important that you have a good sense of what it is that you are trying to measure. Thus, before moving on to summarizing the methods that have been used to identify gentrifying neighborhoods, it is necessary to spend some time discussing how authors in the gentrification literature have answered the seemingly simple question of "what is gentrification?".

It would only be a slight exaggeration to say that there are almost as many definitions of gentrification as there are authors working in the area. However, the one thing that almost all definitions have in common is that gentrification is portrayed as a process in which neighborhoods with predominately lower-socioeconomic status (SES) residents experience an influx of new residents with higher levels of SES. Ruth Glass' original definition referred to "the working-class quarters of

DOI: 10.1201/9781003217459-2

London" being "invaded by the middle-classes". Neil Smith, one of the most important voices in early gentrification research, has described gentrification as "the process … by which poor and working-class neighborhoods in the inner city are refurbished via in influx of private capital and middle-class homebuyers and renters" (Smith (1996)). Finally, in one of the earliest studies to try to utilize quantitative methods to identify gentrifying neighborhoods (Hammel and Wyly (1996)), the authors define gentrification as "the replacement of low-income, inner-city working-class residents by middle- or upper-class households, either through the market for existing housing or demolition to make way for new upscale housing construction". While these definitions differ regarding their specifics, they all identify gentrification as a process in which traditionally lower SES neighborhoods experience an in-migration of residents with higher levels of SES.

Each definition provided in the previous paragraph also identifies gentrification as taking place in inner-city neighborhoods. As the literature has evolved, it has become more common to relax this assumption and to recognize that the type of neighborhood changes associated with gentrification can take place in lower-SES neighborhoods outside of the central city. For example, Markley (2017) studies gentrification in the inner-ring suburbs of Atlanta, while authors such as Phillips (1993) and Lorenzen (2021) have examined rural gentrification. In addition, Lees (2003) identifies "supergentrification" which is described as "the transformation of already gentrified, prosperous and solidly upper-middle-class neighborhoods into much more exclusive and expensive enclaves". While the expansion of gentrification beyond the original definition is clearly of interest to many, this study will use a more traditional definition of gentrification and will focus on measuring gentrification in the inner-city neighborhoods of U.S. cities between 1970 and 2010.

While the general definition of gentrification as lower SES central city neighborhoods experiencing an influx of higher SES residents is straightforward, the literature has increasingly recognized that there are a variety of forms that gentrification can take in a specific neighborhood. One of the more useful typologies of gentrification can be found in Rucks-Ahidiana (2020). While the author defines gentrification quite succinctly as "low-income neighborhoods becoming middle-class", they go on to identify a large variety of forms that gentrification can take. These include *studentification*, which can be thought of as a neighborhood experiencing an in-migration of more highly educated residents who may not have higher incomes than the current residents, and *marginal*, in which white higher-SES residents move into predominantly black and/or Latino low-income neighborhoods. Finally, she also identifies *black* gentrification in which the gentrifying in-movers are black middle-class households. This recognition that gentrification can take many forms is the impetus for analyzing gentrification along multiple dimensions in this study. Specifically, this study will analyze changes in neighborhood income levels, education levels, and the occupations of employed residents to identify gentrifying neighborhoods.

Qualitative vs. Quantitative Approaches

Much of the early gentrification literature can be characterized as more qualitative in nature than quantitative in that the studies typically analyzed how gentrification affected a single neighborhood or a small group of neighborhoods that were simply assumed to have gentrified (Barton (2016)). While the studies often included detailed quantitative analysis of the gentrifying neighborhood or neighborhoods, quantitative methods were not used to determine whether a neighborhood had gentrified.

Quantitative studies, on the other hand, try to use data to identify which of the neighborhoods that could potentially gentrify experience gentrification. As Barton (2016) points out, the standard approach is to employ a "threshold" strategy in which the neighborhoods that could potentially gentrify are identified based on a certain characteristic, and the ones that gentrify are distinguished from those that were not based on their changes in a characteristic or set of characteristics over a certain period. The next section of this chapter will include a more detailed discussion of the characteristics that are often used to identify gentrifiable and gentrifying neighborhoods.

While the advantage of using quantitative methods to identify gentrifying neighborhoods is that the researcher allows the data to determine whether a neighborhood gentrified rather than simply assuming that it is based on personal opinion, one of the most common criticisms of quantitative approaches is that they may not do a good job of separating gentrification from other forms of neighborhood change (Hwang (2015)). In addition, changes in neighborhood SES are only one of the dimensions through which gentrification changes a neighborhood. Quantitative approaches often fail to capture changes to the built environment in the neighborhood (Hwang and Sampson (2014)) or the changes in the local culture (Barton, 2016) brought about by gentrification.

While a purely quantitative approach to identifying gentrifying neighborhoods runs the risk of failing to fully capture certain facets of the gentrification process, it does allow for analyzing gentrification in a large sample of cities and across multiple decades. The hope for the current study is that the benefits of being able to apply a consistent measure of gentrification to both a large sample of cities and across four decades offset the risk that gentrification is not being captured perfectly.

Quantitative Approaches to Identifying Gentrifying Neighborhoods

One thing that quickly becomes clear when attempting to summarize the different quantitative approaches to identifying gentrifying neighborhoods is that there is no single method for identifying such neighborhoods (Barton et al. (2020)). However, as will be seen throughout this section, while the specifics often vary from study to study, there are a lot of similarities in the general approaches. As pointed out by

Hwang and Ding (2020), one of the difficulties in trying to operationalize gentrification is that gentrification can have different characteristics in different cities and at different times.

As summarized in Barton and Cohen (2019), the overall approach in most quantitative studies is to identify "gentrifiable" neighborhoods based on some characteristic at the beginning of a period and then identify "gentrifying" or "gentrified" neighborhoods based on a change in a characteristic or several characteristics over time. This type of approach has become known as a "threshold strategy" (Ding et al. (2016), Barton (2016)). This two-step approach is the most common method used to identify gentrifying neighborhoods for reasons that are very clear. First, for a neighborhood to gentrify, it must be gentrifiable. So, the first thing the researcher must consider is how they are going to determine which neighborhoods are going to be classified as "lower socio-economic status" and, therefore, eligible to gentrify. The second step in the process is to determine the types of changes that will be required for gentrifiable neighborhoods to be classified as "gentrifying" or "gentrified". Generally, the approach is to establish a threshold that the change and/ or changes must exceed in order to identify the gentrifying neighborhoods. Once again, there is variation in the characteristics and thresholds used but the overall approach is very similar across studies.

Finally, before discussing the specifics of how various studies define gentrifiable and gentrifying neighborhoods, it is worth mentioning that virtually all, if not all, quantitative studies of gentrification in the United States use census tracts as a proxy for neighborhoods. This is primarily driven by data availability and, while census tracts are a reasonable approximation of neighborhoods, they often will not line up perfectly with how residents perceive their neighborhoods (Landis (2015)).

Identifying Gentrifiable Neighborhoods

By far, the most common way to identify gentrifiable neighborhoods is to compare the neighborhood income level to the citywide or metropolitan area income. In addition, most quantitative studies only consider central city tracts as candidates for gentrification. One exception is Landis (2015) which considers all tracts within 10 km of the center of the city which, in certain cities, will allow for some suburban neighborhoods to also gentrify.

Two other issues arise when trying to establish the threshold for gentrifiable tracts. First, there is the issue of what is the relevant geography for comparing tract income and, second, at what level to set the threshold. The most common approach in the literature is to identify gentrifiable neighborhoods as central city census tracts with a median household income below the citywide median household income. This definition is used in Hammel and Wyly (1996), Ding et al. (2016), Gibbons and Barton (2016), and Hwang (2020), among others. One potential weakness of this method for identifying gentrifiable neighborhoods is that the threshold is set too high. It is not clear that a neighborhood with an income equal to the citywide median income is necessarily a lower-income neighborhood as is called for in the typical gentrification definition.

There are, however, other thresholds that have been used in the literature. Bostic and Martin (2003) deviate from the most common definition in two ways. First, they calculate tract income relative to metropolitan area income and they also use a much stricter definition and require that a tract's income be less than 50% of the citywide median income to be classified as gentrifiable. This threshold is most likely set too low as it only includes very low-income neighborhoods and excludes low- and moderate-income neighborhoods that should be classified as gentrifiable.

Another common approach is to identify gentrifiable neighborhoods based on their place in the metropolitan distribution of neighborhood incomes or some other measure of SES. For example, Timberlake and Johns-Wolfe (2017) classify a tract as "at risk" of gentrification if it was in the lower three quintiles of a SES scale that they created. This is a more generous threshold than the thresholds mentioned thus far. In another example, Landis (2015) requires that a neighborhood be in the first four deciles of the metropolitan income distribution to be classified as gentrifiable. The author's rationale for using a cutoff below the median is that the four-tenths threshold is roughly comparable to the criteria that the U.S. Department of Housing and Urban Development uses to classify low- and moderate-income neighborhoods. The agency identifies low- and moderate-income neighborhoods as those with incomes below 80% of the area median income. Freeman (2005) also employs a 40th percentile threshold and considers a tract to be gentrifiable if it has an income below the 40th percentile of the metropolitan income distribution. As will be seen later, this definition corresponds very closely to the definition that will be used in this study.

While every example provided so far compares neighborhood income to either city or metropolitan area income, McKinnish et al. (2010) ignore relative income and simply choose all tracts in the bottom quintile of average family income for their sample as their group of potentially gentrifying neighborhoods. Given the wide variation in metropolitan income levels, this approach could potentially exclude lower-income neighborhoods in higher-income metropolitan areas.

Finally, Freeman (2005) uses multiple criteria to identify gentrifiable neighborhoods. The author has three criteria that a neighborhood must meet to be gentrifiable. First, like most of the studies already mentioned, the neighborhood must be in the central city. Second, the neighborhood must have a median income that is either below the median or the 40th percentile of the metropolitan area's median income (both definitions are used). Finally, and this is the unique element of Freeman's definition, the proportion of housing built in the last 20 years in the neighborhood must be below the median (or 40th percentile) of the proportion for the metropolitan area. This third criterion is included to capture the disinvestment that is often part of descriptions of gentrification. While the definition of gentrifiable neighborhoods used in the current study does not include disinvestment, there are three components of Freeman's definition that are similar to how gentrifiable neighborhoods are identified in this study. First, like Freeman, gentrifiable neighborhoods are required to be central city neighborhoods. Second, like Freeman's 40th percentile criteria, the threshold used to identify a lower-income neighborhood is set below the median income. Finally, neighborhood income is calculated relative to the income of the metropolitan area and not the city.

To summarize, the most common way to define gentrifiable neighborhoods in the literature is to define them as central city census tracts with incomes (average or median, family or household) at or below the median income of the city containing the census tract. The current study will deviate from this most common approach in two ways. First, tract income will be compared to metropolitan income rather than city income. More specifically, tract income will be compared to the median income among all tracts in the metropolitan area containing the tract. The rationale for using metropolitan, rather than city income is that, in the author's view, gentrification is a metropolitan-level process. In the same way, the suburbanization represents households and firms choosing more distant locations rather than more centralized locations in a metropolitan area, one of the causes of gentrification is that higher-income households, for whatever reason, are choosing more centralized locations over more distant locations in the metropolitan area.

The second deviation from the more common definition of gentrifiable neighborhoods is that, following Landis (2015), this study will use a stricter threshold to identify the lower-income neighborhoods that could potentially gentrify. The most common threshold requires that a tract only have an income below the median income. This study, in order to mimic the U.S. Department of Housing and Urban Development's system for classifying low- and moderate-income neighborhoods, will consider tracts with incomes that are less than 80% of the median income of all tracts in their metropolitan area as gentrifiable neighborhoods.

Identifying Gentrifying Neighborhoods

Once the pool of gentrifiable or potentially gentrifying tracts has been identified, the next step is to distinguish which of the tracts experience gentrification. As was mentioned above, there is no consensus in the literature as to the best way to identify gentrifying neighborhoods; and a variety of approaches have been employed to identify gentrifying neighborhoods. One convenient way to distinguish the approaches that are used to identify gentrifying neighborhoods is by whether they rely on changes in a single variable (usually income) to identify gentrifying neighborhoods or whether they use multiple variables.

McKinnish et al. (2010) is a good example of a single-variable approach to identifying gentrifying neighborhoods. The authors consider a gentrifiable neighborhood to gentrify if it experienced an increase in average family income between

1990 and 2000 of at least $10,000. Thus, in their approach, a gentrifiable tract gentrifies solely based on the change in its income. While, at the most basic level, gentrification is almost always characterized by lower-income neighborhoods experiencing an increase in their incomes due to the in-migration of higher-income households, there is also much more to gentrification than rising neighborhood income levels. Also, as was the case for how the authors defined gentrifiable neighborhoods, the focus is on absolute changes in income rather than on comparing neighborhood incomes to a local benchmark such as citywide or metropolitan median income. Using changes in absolute income is biased against finding gentrifying neighborhoods in metropolitan areas with lower incomes or in metropolitan areas where the income growth was below $10,000. Since once of the characteristics of gentrifying neighborhoods is that they are moving up distribution of the neighborhood incomes, in relatively low-income metropolitan areas or metropolitan areas with relatively low levels of income growth, a gentrifiable neighborhood could still be moving up the distribution of neighborhood incomes even if its absolute growth is below $10,000.

Landis (2015) is another example of a study that relies on a single variable, income, to identify gentrifying neighborhoods. Landis defines a neighborhood as having undergone "substantial socioeconomic neighborhood change" if it experienced at least a two-decile increase in its place in the metropolitan distribution of neighborhood incomes.

Most of the earlier quantitative work on identifying gentrifying neighborhoods used multiple variables to determine whether a gentrifiable neighborhood experienced gentrification. One of the earliest attempts was Wyly and Hammel (1998). The authors used six socioeconomic variables, three housing variables, and four total population variables to analyze gentrifying neighborhoods. However, the study does not truly use quantitative analysis to distinguish which gentrifiable neighborhoods gentrified. Instead, the authors used a field study to identify neighborhoods that had gentrified, and then determined which of the variables were most closely associated with gentrifying neighborhoods. Because of this, the study never actually establishes quantitative criteria for determining whether a gentrifiable neighborhood gentrified.

Bostic and Martin (2003) used the work by Wyly and Hammel to guide them in a true attempt to use quantitative methods to distinguish gentrifiable neighborhoods that gentrified from those that did not. The authors used nine factors from Hammel and Wyly (1996) and Wyly and Hammel (1999) to identify gentrifying neighborhoods. These nine factors were:

1. The percentage of residents with college degrees at end of the time period
2. Growth in average family income during the time period
3. The homeownership rate at the end of the time period
4. Change in the share of the population aged 30–44 during the time period
5. The poverty rate at the end of the time period

6. The population share of white nonfamily households at end of the time period
7. The Black population share at the end of the time period
8. Managerial and administrative workers as a percentage of the total workforce at end of the time period
9. The percentage of residents with at least some college at the end of the time period.

Once again, however, the authors face the difficulty of how to distinguish between gentrifiable neighborhoods that gentrified and those that did not. Their solution is clever but also quite arbitrary. The authors first employ what they call a *naïve* definition of a gentrifying neighborhood. Using this definition, a gentrifiable neighborhood is deemed to have gentrified if it is gentrifiable at the beginning of a time period and no longer gentrifiable at the end of the time period. While there is some logical appeal to this approach, gentrifiable neighborhoods were defined as central city neighborhoods with incomes less than 50% of the metropolitan median income so, for gentrifiable neighborhoods close to the 50% threshold, relatively low levels of income growth could lead to them being identified as gentrifying. In addition, this criterion suffers from the weakness of identifying gentrifying neighborhood by income growth and nothing else.

While the naïve approach employed by Bostic and Martin suffers from clear weaknesses, its primary purpose was to be a tool that allows the authors to identify gentrifying neighborhoods in their second, multiple-variable approach. In the second approach, each gentrifiable neighborhood in a city was ranked according to each of the nine criteria listed above and the average ranking for each gentrifiable neighborhood was calculated. Then, using the restriction that the same number of neighborhoods would be identified as gentrifying using the second approach as were identified as gentrifying using the naïve approach, the gentrifying neighborhoods were identified. Thus, if the naïve approach identified 15 gentrifiable neighborhoods that were no longer gentrifiable at the end of a time period, then the 15 gentrifiable neighborhoods with the lowest average ranking were identified as gentrifying using the second approach. While this second approach is clearly a multivariable approach, it does not actually establish a non-arbitrary method for identifying gentrifying neighborhoods with quantitative methods.

Freeman (2005) is another good example of a multivariable approach to identifying gentrifying neighborhoods. Freeman requires that a gentrifiable neighborhood experience two changes to gentrify. First, it must experience an increase in its educational attainment that is greater than the median increase for the metropolitan area and, second, it must have an increase in its real housing prices over the period of analysis. One of the things that makes this definition unique is that it does not use neighborhood income change to identify gentrifying neighborhoods. While gentrifying neighborhoods experience many changes other than income growth, one of the changes that is most closely associated with gentrifying neighborhoods is an increase in neighborhood income and it is a curious decision to exclude income from the criteria for identifying gentrifying neighborhoods.

In recent years, it has become more common to rely on multivariable approaches to identify gentrifying neighborhoods. One of the approaches that is most frequently used in the literature was developed by Ding et al. (2016). The authors require a neighborhood to experience an above-citywide median increase in either median gross rent or median home value *and* an above-citywide median increase in educational attainment. Like Freeman, the authors do not use neighborhood income change as one of the criteria for identifying gentrifying neighborhoods. Their rationale for excluding income is that they want to include neighborhoods that experience an increase in young professionals and other groups who might have higher SES but not necessarily higher incomes among their gentrifying neighborhoods.

One of the strengths of the Ding et al. (2016) study is that it allows gentrification to occur over multiple decades. Their study covers the period from 1980 to 2013 and this allows the authors to identify several different types of gentrifying neighborhoods. A tract that gentrified during the 1980s or 1990s was still classified as gentrifiable in 2000, and then gentrified again between 2000 and 2013 was classified as experiencing *continued* gentrification. The tracts that did not gentrify in the 1980s and 1990s but did gentrify after 2000 are classified as experiencing *weak*, *moderate*, or *intense* gentrification based on the magnitude of their changes in the indicators of gentrification.

Gibbons et al. (2018) use the methodology from Ding et al. to identify gentrifying neighborhoods and to identify different types of gentrifying neighborhoods. In addition, the authors add another category of gentrifying neighborhood. Tracts that gentrified during the 1980s and/or 1990s and were no longer gentrifiable in 2000 were classified as *old* gentrification.

Rucks-Ahidiana (2020) also uses a multivariable approach to identify gentrifying neighborhoods but their approach is slightly different than the ones that have been discussed thus far. In order to be gentrified, a gentrifiable tract must have an increase in either the proportion of residents with a college degree *or* an increase in mean family income *and* a percentage change increase in average housing values *or* average rental costs that exceed the increase in the metropolitan area by at least half of a standard deviation. So, while the choice of variables is very similar to the approaches that were discussed above, the bar is set a little bit higher for what it takes to gentrify. Instead of simply requiring that the change in the neighborhood exceeds the change in the city or metropolitan area, the requirement here is for the change in the gentrifiable neighborhood to exceed the metropolitan area change by some additional amount. A similar approach will be used in the current study. Two types of gentrification will be identified. First, if the change in the neighborhood simply exceeds the change in the metropolitan area for the variable of interest, the neighborhood will be said to be *slowly gentrifying*. Second, if the change in the neighborhood exceeds the change in the metropolitan area by 50% or more, the neighborhoods will be classified as *rapidly gentrifying*.

Timberlake and Johns-Wolfe (2017) use an SES scale to identify both gentrifiable and gentrifying tracts. The authors use four variables to construct their index: (1) percentage of residents not in poverty, (2) percentage of residents over the age of 25 with at least a high school degree, (3) percentage employed in professional or technical occupations, and (4) average family income. As was mentioned above, gentrifiable tracts were those in the bottom three quintiles of the SES index distribution while gentrifying tracts were the gentrifiable tracts that moved up at least two quintiles in the distribution between 1980 and 2010.

Candipan (2020) also uses an SES index to identify gentrifying neighborhoods. Gentrifying neighborhoods in this study are those that begin the decade in the bottom four quintiles of the SES index distribution and then experience an increase in their relative SES rank of at least 10 percentage points. Unlike many studies, Candipan also imposes a racial component in identifying gentrifying neighborhoods. Specifically, gentrifying neighborhoods are also required to have above-average growth in their white populations relative to their metropolitan areas.

Barton et al. (2020) is another example of a study that uses a multivariable approach to identify gentrifying neighborhoods with one of the variables imposing an explicitly racial dimension to what is required for a neighborhood to gentrify. In this study, the authors have four requirements for a gentrifiable neighborhood to gentrify. First, the neighborhood must have an increase in the percentage of residents with a bachelor's degree that exceeds the change for the county. Second, the neighborhood must have an increase in median household income that exceeds the change for the county. Third, the neighborhood must have an increase in median gross rent that exceeds the county change. Finally, the explicit racial dimension is that the authors require the neighborhood to have an increase in the percentage of non-Hispanic Whites that exceeds the change for the county.

Gibbons (2021), while not including race as one of the criteria for identifying gentrifying neighborhoods, uses race as a criteria for identifying various types of gentrification. *White gentrification* is a gentrifying neighborhood in which the share of White residents increases and the share of Black, Hispanic, and Asian residents declines. *Non-white gentrification* is a gentrifying neighborhood in which the Black, Hispanic, and Asian population increases but the White population does not. Finally, all other gentrifying neighborhoods are classified as *mixed White gentrification*.

One of the areas where there is a split in the literature is with regard to whether a racial change should be an essential component of identifying gentrifying neighborhoods. As was just mentioned, studies such as Candipan (2020) and Gibbons and Barton (2016) include a racial change variable as part of their definition of gentrifying neighborhoods. However, this is more the exception than the norm. There are far more studies that identify gentrifying neighborhoods without a racial component than studies that include race as a part of the definition. This study will side with the majority of studies and not include race as a variable when identifying gentrifying neighborhoods.

Conclusion

This chapter has summarized the most common approaches to identifying gentrifying neighborhoods using quantitative methods. While a wide variety of approaches have been employed to identify gentrifying neighborhoods, there are some very basic similarities to the various approaches. First, for most studies, neighborhoods are equated with census tracts. Second, almost all studies identify gentrifying neighborhoods with a two-step process. The first step involves identifying neighborhoods that are "eligible" to gentrify, while the second step involves identifying which of the eligible neighborhoods gentrify and which ones do not.

The current study will follow this pattern as well. Census tracts will act as a proxy for neighborhoods, while gentrifiable tracts will be defined as central city census tracts with average household incomes less than 80% of the median average household income of the census tracts in the metropolitan area containing the tract. This definition deviates from the most common definition of gentrifiable tracts in the literature in two ways. First, tract income is compared to metropolitan area income rather than city income and, second, the income threshold is 80% of the median income rather than simply being below the median income. The rationale for using the lower threshold is that gentrifiable tracts should be low- and/or moderate-income neighborhoods and requiring an income at 80% of the median or below lines up with official definitions of low- and moderate-income neighborhoods.

Next, this study does not use one single definition of gentrifying tracts. Instead, Chapters 4, 5, and 6 will each measure a different dimension of gentrification. Chapter 4 will measure the incidence of income gentrification, Chapter 5 will measure educational gentrification, and Chapter 6 will measure occupational gentrification. In each of these chapters, gentrification will be measured in two ways. First, slowly gentrifying tracts will be identified as gentrifiable tracts in which the change in income, education, or occupation exceeds the change at the metropolitan level. Next, rapidly gentrifying tracts will be identified as gentrifiable tracts in which the change in income, education, or occupation exceeds the change at the metropolitan level by at least 50%. The rationale for these two approaches, one that sets the bar rather low and another that sets the bar high, is that it provides a lower bound and an upper bound on the number of gentrifying tracts in each city.

Finally, while Chapters 4–6 will involve measuring gentrification levels in U.S. cities between 1970 and 2010, Chapter 3 will step back and take a look at the trends in metropolitan population and employment in addition to the trends in income, education, and occupation. The goal of Chapter 3 is to establish the socioeconomic environment within which the gentrification activity identified in the later chapters occurred.

Bibliography

Barton, Michael S (2016). "Gentrification and Violent Crime in New York City", *Crime & Delinquency*, 62(9): 1180–1202.

Barton, Michael S. and Isaiah F.A. Cohen (2019). "How Is Gentrification Associated with Changes in the Academic Performance of Neighborhood Schools?", *Social Science Research*, 80(2019): 230–242.

Barton, Michael S., Matthew A. Valasik, Elizabeth Brault and George Tita (2020). "'Gentefiction' in the Barrio: Examining the Relationship Between Gentrification and Homocide in East Los Angeles", *Crime & Delinquency*, 66(13-14): 1888–1913.

Bostic, Raphael W. and Richard W. Martin (2003). "Black Home-Owners as a Gentrifying Force? Neighbourhood Dynamics in the Context of Minority Home-Ownership", *Urban Studies*, 40(12): 2427–2449.

Brown-Saracino, Japonica (2010). *The Gentrification Debates: A Reader*. New York: Routledge.

Brown-Saracino, Japonica (2017). "Explicating Divided Approaches to Gentrification", *Annual Review of Sociology*, 43: 515–530.

Candipan, Jennifer (2020). "Choosing Schools in Changing Places: Examining School Enrollment in Gentrifying Neighborhoods", *Sociology of Education*, 93(3): 215–237.

Ding, Lei, Jackelyn Hwang and Eileen Divringi (2016). "Gentrification and Residential Mobility in Philadelphia", *Regional Science and Urban Economics*, 61(2016): 38–51.

Freeman, Lance (2005). "Displacement or Succession? Residential Mobility in Gentrifying Neighborhoods", *Urban Affairs Review*, 40(4): 463–491.

Gibbons, Joseph (2021). "Measuring Gentrification's Association with Perceived Housing Unaffordability: A Philadelphia Case Study", *Housing Policy Debate*, 31(2): 306–325.

Gibbons, Joseph and Michael S. Barton (2016). "The Association of Minority Self-Rated Health with Black versus White Gentrification", *Journal of Urban Health*, 93(6): 909–922.

Gibbons, Joseph, Michael S. Barton and Elizabeth Brault (2018). "Evaluating Gentrification's Relation to Neighborhood and City Health", *PLoS ONE*, 13(11): 1–18.

Glass, Ruth (1964). *London: Aspects of Change*. London: MacGibbon & Kee.

Hammel, Daniel J. and Elvin K. Wyly (1996). "A Model for Identifying Gentrified Areas with Census Data", *Urban Georgraphy*, 17(3): 248–268.

Hwang, Jackelyn (2015). "Gentrification in Changing Cities: Immigration, New Diversity, and Racial Inequality in Neighborhood Renewal", *Annuals, AAPSS*, 660: 319–340.

Hwang, Jackelyn (2020). "Gentrification Without Segregation? Race, Immigration, and Renewal in a Diversifying City", *City and Community*, 19(3): 538–572.

Hwang, Jackelyn and Lei Ding (2020). "Unequal Displacement: Gentrification, Racial Stratification, and Residential Destinations in Philadelphia", *American Journal of Sociology*, 126(2): 354–406.

Hwang, Jackelyn and Robert Sampson (2014). "Divergent Pathways of Gentrification: Racial Inequality and the Social Order of Renewal in Chicago Neighborhoods", *American Sociological Review*, 79(4): 726–751.

Landis, John D. (2015) "Tracking and Explaining Neighborhood Socioeconomic Change in U.S. Metropolitan Areas Between 1990 and 2010", *Housing Policy Debate*, pp. 1–51.

Lees, Loretta (2003). "Super-Gentrification: The Case of Brooklyn Heights, New York City", *Urban Studies*, 40(12): 2487–2509.

Lees, Loretta, Tom Slater and Elvin Wyly (2008). *Gentrification*. New York: Routledge.

Lorenzen, Matthew (2021). "Rural Gentrification, Touristification, and Displacement: Analyzing Evidence from Mexico", *Journal of Rural Studies*, 86: 62–75.

Markley, Scott (2017). "Suburban Gentrification? Examining the Geographies of New Urbanism in Atlanta's Inner Suburbs", *Urban Geography*, 39(4): 606–630.

McKinnish, Terra, Randall Walsh and Kirk T. White (2010). "Who Gentrifies Low-Income Neighborhoods?", *Journal of Urban Economics*, 67(2): 180–193.

Phillips, Martin (1993). "Rural Gentrification and the Processes of Class Colonization", *Journal of Rural Studies*, 9(2): 123–140.

Rucks-Ahidiana, Zawadi (2020). "Racial Composition and Trajectories of Gentrification in the United States", *Urban Studies*, 587(13): 2721–2741.

Smith, Neil (1996). *The New Urban Frontier: Gentrification and the Revanchist City*. New York: Routledge.

Timberlake, Jeffrey M. and Elaina Johns-Wolfe (2017). "Neighborhood Ethnoracial Composition and Gentrification in Chicago and New York, 1980 to 2010", *Urban Affairs Review*, 53(2): 236–272.

Wyly, Elvin K. and Daniel J. Hammel (1999). "Modeling the Context and Contingency of Gentrification", *Journal of Urban Affairs*, 20(3): 303–326.

3

METROPOLITAN AND CENTRAL CITY TRENDS IN INCOME, EDUCATION, AND OCCUPATION

Introduction

The next three chapters will calculate the amount of income, educational, and occupational gentrification in a sample of U.S. cities from 1970 to 2010. However, before moving on to the process of identifying gentrifying neighborhoods, the current chapter will analyze a variety of economic and demographic trends in U.S. metropolitan areas to capture the overall metropolitan-level changes that were occurring while neighborhoods were gentrifying.

The analysis in this chapter will focus mainly on answering two questions. First, how has the central city share of a variety of variables (population, employment, income, educational attainment, workers employed in professional and executive occupations) changed over time? The rationale for looking at changes in the central city share of these variables is straight-forward. We would expect to see gentrification levels rising if the central city share of college graduates, for example, is increasing. The second question is how does central city growth in variables such as population, income, residents with a college degree, and workers employed in professional and executive occupations compare with growth in these variables in the rest of the metropolitan area? Once again, the expectation is that there will be more gentrification in cities that are experiencing higher levels of growth than the rest of the metropolitan areas. For each variable studied, the general trends for the whole sample will be identified as well as the cities with the highest and lowest values for each variable. This aids in identifying the cities that are most likely to experience higher levels of gentrification and those in which gentrification is less likely.

The analysis in this chapter makes use of the State of the Cities Data System (SOCDS) which is available from the U.S. Department of Housing and Urban Development's Office of Policy Development and Research. The data found in the

DOI: 10.1201/9781003217459-3

SOCDS is very useful because it provides data on many socioeconomic variables at a variety of geographic levels. These levels include cities, suburbs, and metropolitan areas. This makes it ideal for comparing the experiences of central cities to those of other parts of the metropolitan areas.

Metropolitan Population and Employment Trends

Table 3.1 contains the basic information regarding the central city share of population and employment for the cities that are included in the sample used in this study. This sample includes the 100 cities with the largest populations in 1970. The SOCDS is used to identify the total population and total employment in each of the cities as well as the total population and total employment of the metropolitan area containing the cities for 1970, 1980, 1990, 2000, and 2010.[1] Then, the share of the total metropolitan population and employment contained in the cities in the sample is calculated for each year.

In 1970, the cities included in the sample contained 20.2% of the population and 20.7% of the total employment in the metropolitan areas included in the study. El Paso had the highest metropolitan population share, with 89.7% of the metropolitan area's population, while Paterson, NJ, with 0.9%, had the smallest metropolitan population share. Unsurprisingly, given the size of the metropolitan area, the three lowest metropolitan population shares were for cities in the New York metropolitan area. Jacksonville, FL, had the highest share of metropolitan employment with 87.2% of the total employment, while Paterson also had the lowest share of metropolitan employment with 0.8%.

Given the well-known trend of post-1950s suburbanization in U.S. cities (Mieszkowski and Mills (1993)), it is not surprising that the central city share of both population and employment decreased over time. The largest decrease is in the 1970s when the central city share of the population fell by 1.7 percentage points and the central city share of employment fell by 2.5 percentage points. The population share fell an additional 1.0 percentage points in the 1980s, 0.7 percentage points in the 1990s, and 0.7 percentage points in the 2000s so that, by 2010, the central city population share was 16.2%. The central city share of metropolitan employment fell by 1.4 percentage points in the 1980s, 0.5 percentage points in the 1990s, and 0.5 percentage points in the 2000s and was 15.8% in 2010. In 2010, Lincoln, NE, had both the highest metropolitan population and employment shares with 84.3% of the metropolitan population and 82.9% of the metropolitan employment. Paterson still had the lowest metropolitan population share with 0.8% of the metropolitan population, while Gary, IN, had the lowest employment share at 0.6%. The results in Table 3.1 are important because they highlight that the gentrification trends that will be identified in the next three chapters were not a result of a reversal of population and employment suburbanization. While the rate of suburbanization was decreasing slightly, the central city share of metropolitan population and employment was uniformly decreasing during the timeframe included in this study.

TABLE 3.1 Central City Share of Metropolitan Population and Employment (Full Sample)

	Central City % of Metropolitan Area Total					Change in Central City Share (Percentage Points)			
	1970	1980	1990	2000	2010	1970s	1980s	1990s	2000s
Population	20.2%	18.6%	17.5%	16.8%	16.2%	−1.68	−1.01	−0.74	−0.66
Highest	El Paso	El Paso	El Paso	Lincoln	Lincoln	San Jose	Fresno	Fort Wayne	Wichita
	89.7%	88.6%	87.1%	84.6%	84.3%	6.5	10.7	3.9	1.3
Lowest	Paterson	Paterson	Paterson	Paterson	Paterson	Tucson	Memphis	Jackson, MS	New Orleans
	0.9%	0.8%	0.8%	0.8%	0.8%	−12.6	−7.6	−6.9	−15.0
Employment	20.7%	18.2%	16.8%	16.3%	15.8%	−2.48	−1.42	−0.46	−0.51
Highest	Jacksonville	El Paso	El Paso	Lincoln	Lincoln	Columbus, GA	Fresno	Fort Wayne	New York
	87.2%	88.2%	87.9%	85.1%	82.9%	7.3	8.0	3.3	1.8
Lowest	Paterson	Paterson	Paterson	Paterson	Gary, IN	Denver	Memphis	Jackson, MS	New Orleans
	0.8%	0.7%	0.8%	0.6%	0.6%	−13.9	−10.3	−8.9	−13.2

TABLE 3.2 Total Change in Central City Share of Population
and Employment, 1970–2010 (Full Sample)

Population	1970–2010
Largest	
Fresno, CA	13.36
Virginia Beach, VA	10.97
San Jose, CA	10.18
Louisville, KY	9.33
Charlotte, NC	8.42
Smallest	
New Orleans, LA	−30.98
Denver, CO	−22.58
Jacksonville, FL	−22.29
Nashville, TN	−21.69
Phoenix, AZ	−20.78
Employment	
Largest	
San Jose, CA	13.04
Virginia Beach, VA	11.47
Fresno, CA	10.52
Columbus, GA	9.92
Charlotte, NC	9.77
Smallest	
New Orleans, LA	−32.95
Jacksonville, FL	−25.16
Denver, CO	−24.96
Memphis, TN	−23.70
Nashville, TN	−23.37

Table 3.1 also contains the cities that experienced the largest increases and decreases in population and employment share in each decade, while Table 3.2 contains the cities that experienced the largest increases and decreases over the entire 1970–2010 period. The results for the 1970s in Table 3.1 reveal that San Jose had the largest increase in its share of the metropolitan population during the 1970s with an increase of 6.5 percentage points, while Columbus, GA, with an increase of 7.3 percentage points, had the largest increase in metropolitan employment share. There were only ten cities in which the central city share of the metropolitan population increased during the 1970s, while there were 13 cities in which the metropolitan employment share increased.

Fresno, CA, had the largest increase in both metropolitan population share and metropolitan employment share during the 1980s. Fresno's metropolitan population share increased by 10.7 percentage points while its employment share increased by 8.0 percentage points. During the 1980s, there were 17 cities in which the metropolitan population share increased as well as 17 cities in which the metropolitan employment share increased.

Fort Wayne, IN, had the largest increase in both metropolitan population share and metropolitan employment share in the 1990s with a 3.9 percentage points increase in population share and a 3.3 percentage points increase in employment share. During the 1990s, there were 14 cities in which the metropolitan population share increased and 11 cities in which the metropolitan employment share increased.

Louisville, KY, and Fort Wayne had the largest increases in both metropolitan population and employment share in the 2000s. However, this is misleading. In 2003, the city of Louisville merged with Jefferson County and created a single governmental entity and, in 2006, Fort Wayne annexed over 12 square miles of surrounding areas. Thus, much of the increase in population and employment share in both cities in the 2000s was due to geographical expansion rather than growth. Because of this, both cities will be ignored when calculating the largest increases for the 2000s through the rest of the chapter. Ignoring Louisville and Fort Wayne, Wichita had the largest increase in metropolitan population share, with an increase of 1.3 percentage points, while New York had the largest increase in metropolitan employment share with a 1.8 percentage points increase. During the 2000s, there were nine cities in which the metropolitan population share increased and 23 cities in which the metropolitan employment share increased.

Table 3.2 contains the cities with the largest cumulative changes from 1970 to 2010. Fresno had the largest increase in metropolitan population share with an increase of 13.4 percentage points, while San Jose had the largest increase in metropolitan employment share with an increase of 13.0 percentage points. In total, there were only 11 cities in which their share of metropolitan population increased from 1970 to 2010 and ten cities in which the employment share increased.

Returning to Table 3.1, in the 1970s, Tucson, AZ, had the largest decrease in its metropolitan population share with a decrease of 12.6 percentage points. Denver had the largest decrease in metropolitan employment share with a 13.9 percentage points decrease. In the 1970s, there were four cities (Tucson, Denver, Jacksonville, and Salt Lake City) with decreases in metropolitan population share greater than 10 percentage points and 13 cities with decreases in their metropolitan employment share that exceeded 10 percentage points.

During the 1980s, Memphis had the largest decreases in both metropolitan population and employment share with a 7.6 percentage points decrease in population share and a 10.3 percentage points decrease in employment share. There were no cities in the 1980s with declines in metropolitan population share above 10 percentage points. However, there were six cities (Memphis, Houston, Denver, Atlanta, Phoenix, and Jackson (MS)) with declines that exceeded 5 percentage points. Memphis was the only city in which the decline in metropolitan employment share exceeded 10 percentage points. There were an additional 12 cities in which the metropolitan employment share fell by more than 5 percentage points.

Jackson, MS, had the largest declines in both metropolitan population and employment shares in the 1990s with a 6.9 percentage point decrease in population share and an 8.9 percentage point decrease in employment share. While there were no cities with double-digit decreases in either population or employment share, there were five cities (Jackson, Baltimore, Indianapolis, Des Moines, and Nashville) with decreases in population share that exceeded 5 percentage points and six cities (Jackson, Nashville, Des Moines, Indianapolis, Baltimore, and Birmingham) in which the employment share fell by more than 5 percentage points.

During the 2000s, New Orleans had the largest decreases in both population and employment shares with a 15.0 percentage point decrease in population share and a 13.2 percentage points decrease in employment share. However, both results were elevated by the impact of Hurricane Katrina. Phoenix had the next largest decrease in population share with a 5.2 percentage points decrease, while Indianapolis, with a 5.6 percentage point decrease, had the second largest decline in employment share. New Orleans was the only city in the 2000s with a decline in either population or employment share that exceeded 10 percentage points. There were two cities (Phoenix and Rockford) with decreases in population share above 5 percentage points, while Indianapolis was the only other city with a decline in employment share above 5 percentage points.

Returning to Table 3.2, New Orleans had the largest decreases in both metropolitan population and employment shares between 1970 and 2010. New Orleans' share of the metropolitan population decreased by 31.0 percentage points while its share of metropolitan employment fell by 33.0 points. Denver had the second-largest decrease in metropolitan population share, with a 22.6 percentage point decrease, while Jacksonville had the second-largest overall decline in employment share with a 25.2 percentage point decrease. In all, there were five cities (New Orleans, Denver, Jacksonville, Nashville, and Phoenix) in which the metropolitan population share decreased by more than 20 percentage points between 1970 and 2010. There were also 16 cities in which the metropolitan employment share decreased by at least 20 percentage points.

Table 3.3 shifts the focus away from the central city share of population and employment to population and employment growth rates. For each decade, the table contains the average population and employment growth for metropolitan areas and cities in the sample as well as the metropolitan areas and cities with the highest and lowest population and employment growth. In the 1970s, the average metropolitan area in the sample experienced population growth of 5.6% and employment growth of 19.7%. The average city in the sample lost 3.1% of its population and had an employment growth of 5.3%. Thus, there was almost a 9-percentage point difference between metropolitan and city population growth while metropolitan employment growth was more than 14 percentage points greater than city employment growth. Among metropolitan areas, Phoenix had the highest population growth at 54.1% and Tucson had the highest employment growth at 80.8%, while

TABLE 3.3 Population and Employment Growth

Growth Rates by Decade

Population Growth

		1970s	1980s	1990s	2000s	
	Metropolitan					
	Areas	5.63%	10.47%	12.16%	4.30%	
	Central Cities	−3.12%	4.43%	7.46%	0.20%	
Highest		1970s	1980s	1990s	2000s	Total
	Metropolitan	Phoenix	Riverside	Austin	Phoenix	Phoenix
	Areas	54.5%	66.1%	47.7%	24.2%	290%
	Central Cities	Virginia Beach	Fresno	Austin	Charlotte	Fresno
		52.4%	62.3%	41.0%	19.9%	188%
Lowest						
	Metropolitan	Buffalo	Youngstown	Youngstown	New Orleans	Buffalo
	Areas	−7.9%	−7.0%	−1.7%	−22.2%	−15.7%
	Central Cities	St. Louis	Gary	Youngstown	New Orleans	New Orleans
		−27.2%	−23.2%	−14.3%	−53.9%	−62.4%

Employment Growth

		1970s	1980s	1990s	2000s	
	Metropolitan					
	Areas	19.67%	19.02%	6.36%	9.76%	
	Central Cities	5.31%	9.80%	3.33%	6.37%	
Highest		1970s	1980s	1990s	2000s	Total
	Metropolitan	Tucson	Riverside	Austin	Riverside	Austin
	Areas	80.8%	75.0%	51.3%	36.6%	392%
	Central Cities	San Jose	Virginia Beach	Austin	Riverside	Austin
		89.9%	65.7%	45.4%	36.6%	272%
Lowest						
	Metropolitan	Buffalo	Shreveport	Hartford	New Orleans	Buffalo
	Areas	1.2%	−4.7%	−4.4%	−20.6%	3.4%
	Central Cities	Detroit	Gary	Hartford	New Orleans	New Orleans
		−29.7%	−24.0%	−25.5%	−51.9%	−55.7%

Buffalo experienced both the largest population losses with a loss of 7.9% of its population and the lowest level of employment growth at 1.2%. Among cities, Virginia Beach experienced the highest population growth at 52.4%, while San Jose experienced the highest employment growth at 89.9%. St. Louis had the largest population loss at 27.2%, while Detroit experienced the highest employment losses at 29.7%.

In the 1980s, the average metropolitan area in the sample experienced 10.5% population growth and 19.0% employment growth while the average city experienced 4.4% population growth and 9.8% employment growth. Thus, the growth differentials were smaller in the 1980s than in the 1970s. The population differential decreased from 8.8 percentage points to 6.0 percentage points while the employment differential fell from 14.4 percentage points to 9.2 percentage points. Among metropolitan areas, Riverside had the highest levels of both population and employment growth with 66.1% population growth and 75.0% employment growth. Youngstown had the largest population losses among metropolitan areas with a population decline of 7.0%, while Shreveport had the largest employment decline at 4.7%. Among cities, Fresno, with a growth of 62.3%, had the highest level of population growth and Virginia Beach, with 65.7% growth, had the highest level of employment growth. Gary, IN, had both the largest population losses and the largest employment losses among cities with a decline of 23.2% in its population and 24.0% in its total employment.

In the 1990s, the growth differentials between central cities and metropolitan areas narrowed again. The average metropolitan area had 12.2% population growth and 6.4% employment growth while the average city had a population growth of 7.5% and employment growth of 3.3%. Thus, the population growth gap decreased from 6.0 points in the 1980s to 4.7 points and the employment growth gap decreased from 9.2 points in the 1980s to 3.0 points. Among metropolitan areas, Austin had the highest levels of both population and employment growth with a population growth of 47.7% and employment growth of 51.3%. Youngstown, once again, had the largest population loss with a loss of 1.7% of its population, while Hartford had the largest employment loss with a 4.4% decrease in total employment. Among cities, Austin also had the highest levels of both population and employment growth. During the 1990s, the city of Austin's population grew by 41.0% while its total employment grew by 45.4%. Youngstown also had the largest population losses among the cities in the sample with a 14.3% decrease in its population. Hartford had the largest employment losses with a decrease of 25.5% in its total employment.

Finally, in the 2000s, the average metropolitan area in the sample had a population growth of 4.3% and employment growth of 4.1%. The average city in the sample had 0.2% population growth and 3.4% employment growth. Thus, the population growth gap narrowed once again from 4.7 points in the 1990s to 4.1 points while the employment growth gap increased slightly from 3.0 points to 3.4 points. Among metropolitan areas, Phoenix had the highest population growth in the 2000s with an increase of 24.2%, while Riverside had the highest employment growth with an increase of 36.6%. The New Orleans metropolitan area had the largest losses in both population and employment in the 2000s with a 22.2% decrease in population and a 20.6% decrease in employment. However, as was pointed out earlier, the losses in New Orleans during the 2000s are elevated due to Hurricane Katrina. After New Orleans, Youngstown had the largest population losses among metropolitan areas with a 5.6% decrease in population. Flint had the largest decrease in total employment in the 2000s with a 3.2% decrease. Among cities, Charlotte had the highest population growth in the 2000s with a 19.9% increase, while

Riverside had the largest increase in total employment with an increase of 36.6%. New Orleans also had the largest population and employment losses among cities with a 53.9% decrease in population and a 51.9% decrease in total employment. After New Orleans, Youngstown had the largest decrease in population of 19.7% and Detroit had the largest decrease in employment with a 19.4% decrease in total employment.

The final column of Table 3.3 contains the results regarding the total growth from 1970 to 2010. Phoenix led all metropolitan areas in population growth with 290% growth and was followed by Austin with 278% growth. Austin had the highest cumulative employment growth among metropolitan areas with 392% growth and was followed closely by Phoenix with 389% employment growth. Buffalo had the largest population losses among metropolitan areas with a loss of 15.7% of its population. Buffalo also had the smallest employment growth with 3.4% cumulative growth. Among cities, Fresno had the highest population growth with 188% growth. Austin had the next highest cumulative population growth with 185% growth. Austin led all cities in employment growth with cumulative growth in total employment of 272%. New Orleans suffered the largest population losses among cities with a loss of 62.4% of its population. The next highest population loss among cities was for Youngstown with a 52.9% loss. New Orleans also had the largest decrease in total employment with a decrease of 55.7%. After New Orleans, the largest employment loss among cities was for Gary with a 53.6% decrease.

Once again, the results in Table 3.3 show that the gentrification trends that are identified in the next three chapters occurred at a time when suburban growth in population and employment exceeded the growth in central cities. However, the growth rate differential between metropolitan areas and central cities, for both population and employment, decreased substantially between the 1970s and the 2000s.

Metropolitan Income Trends

The next three chapters will measure the amount of income, educational, and occupational gentrification in U.S. cities from 1970 to 2010. Before moving on to that analysis, the rest of this chapter will focus on metropolitan trends in income, educational attainment, and occupation. As with the previous analysis regarding population and employment, the goal will be to compare how central cities and metropolitan areas differ with respect to how their incomes, educational attainment, and occupational mix changed between 1970 and 2010. In addition, the cities with the largest and smallest changes in these characteristics will be identified.

Median household incomes for all central cities in the sample, as well as the metropolitan areas they belong to, were collected from the State of the Cities Database for 1970, 1980, 1990, 2000, and 2010. Table 3.4 reports the average values for both geographies for each year. The table also reports an average central city income as a percentage of the average metropolitan area income for each year. Table 3.5 reports the cities with the highest and lowest incomes relative to their

TABLE 3.4 Median Household Income for Sample

	Average MHI for Sample					% Change in Average MHI				
	1970	1980	1990	2000	2010	1970s	1980s	1990s	2000s	Total
Central Cities	$41,724	$43,880	$44,516	$45,647	$42,491	5.17%	1.45%	2.54%	−6.91%	1.84%
Metropolitan Areas	$46,984	$51,481	$54,512	$57,226	$54,042	9.57%	5.89%	4.98%	−5.56%	15.02%
CC as % of MSA	88.80%	85.24%	81.66%	79.77%	78.63%	−3.57%	−3.57%	−1.89%	−1.14%	−10.18%

TABLE 3.5 Central City Relative Incomes

Central City Income as a % of Metropolitan Area Income

	1970	1980	1990	2000	2010
Highest	Columbus, GA 128%	Virginia Beach 124%	Virginia Beach 118%	Virginia Beach 115%	Virginia Beach 116%
Lowest	Hartford 64%	Hartford 57%	Hartford 53%	Hartford 47%	Bridgeport 47%

Change in Central City Income as a % of Metropolitan Income

	1970s	1980s	1990s	2000s	Total
Highest	Madison 20.2	Boston 4.2	San Francisco 8.1	Fort Lauderdale 10.5	San Francisco 16.1
Lowest	Columbus, GA −23.5	Gary −30.1	Warren −14.4	Youngstown −12.1	Gary −38.5

metropolitan area's income as well as the cities with the largest changes in relative income in each decade and over the entire 1970–2010 period.

In 1970, the average central city income was 88.8% of the average metropolitan income. Columbus, GA, had the highest relative central city income with the city having a median income that was 27% higher than the metropolitan area's median income. The lowest value was for Hartford, CT, where the central city median income was 36% below the metropolitan area's median income.

During the 1970s, the average central city median income increased by 5.2% while the average metropolitan median income increased by 9.6%. Because of this, the income ratio fell by 3.6 percentage points to 85.2%. The largest increase in the income ratio was for Madison, WI, where the ratio increased by 20.2 percentage points. The largest decrease was for Columbus, GA, which had the highest ratio in 1970. The ratio in Columbus fell by 23.5 percentage points in the 1970s. In 1980, Virginia Beach, with a ratio of 124.4%, had the highest value, while Hartford, whose ratio decreased to 57.4%, had the lowest value. In the 1970s, there were 21 cities where the income ratio increased. This, of course, means that there were 79 cities in which the central city income growth was below that of the metropolitan area.

During the 1980s, the average central city median income increased by 1.5% while the average metropolitan median income increased by 5.9%. The income ratio decreased by an additional 3.6 points to 81.7% in 1990. The largest increase in the income ratio was for Boston where the ratio increased by 4.2 percentage points in the 1980s, while the largest decrease in the income ratio was for Gary, IN, where the income ratio fell by 30.1 percentage points. In 1990, the highest value for the income ratio continued to be for Virginia Beach which had a ratio of 118.0%, while Hartford, with a value of 53.4%, continued to have the lowest value. During the 1980s, there were 17 cities in which the income ratio increased and 83 cities in which central city income growth lagged metropolitan income growth.

In the 1990s, the average central city median income increased by 2.5% while the average metropolitan median income increased by 5.0%. Thus, the income ratio once again decreased, this time by 1.9 percentage points. San Francisco had the largest increase in its income ratio with an increase of 8.1 percentage points, while Warren, MI, had the largest decrease in its ratio, with a decline of 14.4 percentage points. In 2000, Virginia Beach, with an income ratio of 114.8%, continued to have the highest value, while Hartford, with an income ratio of 47.2%, continued to have the lowest ratio. During the 1990s, there were 27 cities in which the income ratio increased and 73 cities where it decreased.

During the 2000s, the average central city median income decreased by 6.9% while the average metropolitan median income decreased by 5.6%. Thus, the income ratio continued to fall, this time by 1.1 percentage points. The largest increase in the income ratio during the 2000s was for Fort Lauderdale, FL, which had a 10.5 percentage point increase, while the largest decrease was for Youngstown where the income ratio decreased by 12.1 percentage points during the 2000s. In 2010, Virginia Beach, with an income ratio of 115.8%, continued to have the highest income ratio, while Bridgeport, CT, with an income ratio of 46.6%, replaced Hartford as the city with the lowest income ratio. During the 2000s, there were 34 cities in which the income ratio increased and 66 cities where it decreased.

Between 1970 and 2010, the average median city income increased by 1.8% while the median metropolitan income increased by 15.0%. Because of this, the income ratio fell by a total of 10.2 points between 1970 and 2010. San Francisco had the largest cumulative increase in its income ratio with a total increase of 16.1 percentage points, while Gary, IN, had the largest cumulative decrease with a decline of 38.5 percentage points. In total, there were 14 cities in which the income ratio increased between 1970 and 2010 and 86 cities in which the ratio decreased.

The results in Tables 3.4 and 3.5 show that the period from 1970 to 2010 was one where central city income growth was lower than the income growth in their metropolitan areas. However, the gap between central city income growth and metropolitan income growth was greatest in the 1970s and 1980s, with over 70% of the decline in the income ratio occurring between 1970 and 1990. While income ratios were falling in most cities, with only 14 cities having an increase between 1970 and 2010, there was an increasing number of cities with rising income ratios over time. In the 1970s, there were 21 cities in which the income ratio increased. This number fell to 17 cities in the 1980s but increased to 27 cities in the 1990s and 34 cities in the 2000s. Thus, there were a substantial number of cities in which central city income growth exceeded suburban income growth which means that these cities are likely candidates to have experienced income gentrification.

Table 3.6 shows the cities and metropolitan areas with the highest and lowest income growth in each decade and over the entire 1970–2010 period. In the 1970s, Austin, TX, had the highest income growth among metropolitan areas with 52.1% growth, while Madison, WI, had the highest growth among cities with 54.4% growth. The metropolitan area with the lowest income growth was Dayton, OH, where median household income fell by 5.5%. Among cities, Newark, NJ, had the largest decline in income with a 17.4% decrease. During the 1970s, there were eight metropolitan areas and 30 cities in which the median household income decreased.

During the 1980s, the highest income growth among metropolitan areas was for Bridgeport, CT, with 33.1% growth, while Boston had the highest income growth among cities with 36.3% growth. The largest income decrease among metropolitan areas during the 1980s was for Youngstown, OH, where median household income decreased by 16.6%, while Gary, IN, had the largest decrease among cities with a 34.1% decrease. During the 1980s, there were 35 metropolitan areas and 48 cities in which the median household income decreased. This was a very large increase over the eight metropolitan areas and 30 cities that had declining income in the 1970s.

During the 1990s, Austin once again had the largest increase in income among metropolitan areas with a 30.3% increase in median household income. Austin also had the highest income growth among cities with a 25.0% increase. The largest decrease in income among metropolitan areas was for Los Angeles which had a 7.5% decrease in income. Among cities, the largest decrease was for Hartford, CT,

TABLE 3.6 Metropolitan and Central City Income Growth

Panel A: Metropolitan Areas

Highest	*1970s*		*1980s*		*1990s*		*2000s*		*Total*	
	MSA	% Change	MSA	% Change	MSA	% Change	MSA	% Change	MSA	% Change
	Austin, TX	52.1%	Bridgeport, CT	33.1%	Austin, TX	30.3%	New Orleans, LA	8.7%	Austin, TX	87.2%
	Lincoln, NE	36.0%	Boston, MA	28.5%	Salt Lake City, UT	19.8%	San Diego, CA	4.6%	San Diego, CA	64.5%
	Baton Rouge, LA	32.0%	New York, NY	27.0%	Denver, CO	16.0%	Riverside, CA	3.8%	Norfolk, VA	51.9%

Lowest	*1970s*		*1980s*		*1990s*		*2000s*		*Total*	
	MSA	% Change	MSA	% Change	MSA	% Change	MSA	% Change	MSA	% Change
	Dayton, OH	-5.5%	Youngstown, OH	-16.6%	Los Angeles, CA	-7.5%	Flint, MI	-17.7%	Flint, MI	-23.3%
	New York NY	-3.4%	Flint, MI	-12.9%	Hartford, CT	-5.5%	Rockford, IL	-16.3%	Youngstown, OH	-18.8%
	Buffalo, NY	-1.7%	Pittsburgh, PA	-11.6%	Riverside, CA	-5.2%	Greensboro, NC	-15.4%	Dayton, OH	-14.9%

(Continued)

TABLE 3.6 (Continued)

Panel B: Cities

	1970s City	% Change	1980s City	% Change	1990s City	% Change	2000s City	% Change	Total City	% Change
Highest	Madison, WI	54.4%	Boston, MA	36.3%	Austin, TX	25.0%	Louisville, KY	19.5%	San Francisco, CA	76.3%
	Norfolk, VA	36.0%	Jersey City NJ	33.0%	San Francisco, CA	23.0%	Arlington, VA	14.6%	San Diego, CA	72.0%
	Lincoln, NE	35.5%	Paterson, NJ	31.5%	Salt Lake City, UT	21.2%	New Orleans, LA	9.2%	Madison, WI	69.0%
	1970s City	% Change	1980s City	% Change	1990s City	% Change	2000s City	% Change	Total City	% Change
Lowest	Newark, NJ	−17.4%	Gary, IN	−34.1%	Hartford, CT	−16.6%	Youngstown, OH	−25.4%	Youngstown, OH	−47.0%
	Dayton, OH	−15.1%	Flint, MI	−31.2%	Long Beach, CA	−13.1%	Rockford, IL	−24.2%	Gary, IN	−41.1%
	Jersey City NJ	−14.6%	Youngstown, OH	−25.7%	Syracuse, NY	−12.4%	Grand Rapids, MI	−23.0%	Flint, MI	−40.7%

where median household income decreased by 16.6% during the 1990s. There were 19 metropolitan areas and 30 cities in which median household income decreased during the 1990s which represented a substantial improvement over the 35 metropolitan areas and 48 cities with declining incomes in the 2000s.

During the 2000s, the highest income growth among metropolitan areas was for New Orleans which had an 8.7% increase in median household income. For cities, the highest income growth was for Arlington, VA, which had a 14.6% increase in income. The largest decline in income among metropolitan areas was for Flint, MI, which saw median household income decrease by 17.1%. For cities, the largest decrease was for Youngstown where median household income decreased by 25.4%. In total, there were 87 metropolitan areas and 78 cities where the median household income decreased. Interestingly, this is the only decade in which more metropolitan areas than cities had declining incomes. This could be a sign that cities weathered the economic turmoil of the 2000s better than their surrounding metropolitan areas and suggests that there could be higher levels of income gentrification in the 2000s than in other decades.

The final column of Table 3.6 shows the metropolitan areas and cities with the largest cumulative changes in income from 1970 to 2010. Among metropolitan areas, Austin, TX, had the largest cumulative increase in income with an 87.2% increase. For cities, the largest increase was for San Francisco where the median household increased by 76.3% from 1970 to 2010. The largest decrease in income among metropolitan areas was for Flint, MI, where median household income decreased by 23.3% from 1970 to 2010. Among cities, the largest cumulative decrease in income was for Youngstown, OH, where income fell by 47.0%. From 1970 to 2010, there were 18 metropolitan areas and 48 cities in which the median household income decreased.

The results in Table 3.6 reveal that, in general, income growth from 1970 to 2010 was much greater in the suburban portions of metropolitan areas than in the central cities. However, the gap between the suburban and central city growth rates narrowed over time and there was an increasing number of cities in which income growth exceeded metropolitan income growth. This suggests that income gentrification could become more common over time as more central city areas experienced higher levels of income growth.

Metropolitan Educational Trends

Table 3.7 compares the levels of educational attainment between metropolitan areas and central cities. Educational attainment is measured as the percentage of residents aged 25 and above with at least a bachelor's degree. As with the previous tables, the average levels of educational attainment for the metropolitan areas and central cities in the sample are reported, as well as the metropolitan areas and cities with the highest and lowest levels of educational attainment. The table also reports the areas with the highest and lowest changes in educational attainment for each decade and for the entire 1970–2010 period.

TABLE 3.7 Educational Attainment for Cities and Metropolitan Areas

Panel A: Percentage of Residents 25 and Over With At Least a Bachelor's Degree

	1970	1980	1990	2000	2010
Metropolitan Areas	11.9%	17.9%	22.2%	26.5%	29.2%
Highest	Washington, DC 22.1%	Washington, DC 31.2%	Washington, DC 37.5%	Washington, DC 42.5%	Washington, DC 46.1%
Lowest	Youngstown, OH 7.1%	Youngstown, OH 10.7%	Flint, MI 12.8%	Flint, MI 16.2%	Youngstown, OH 17.4%
Cities	11.4%	17.1%	21.1%	24.5%	27.1%
Highest	Arlington, VA 29.6%	Arlington, VA 42.5%	Arlington, VA 52.3%	Arlington, VA 60.2%	Arlington, VA 67.2%
Lowest	Newark, NJ 4.3%	Paterson, NJ 6.2%	Cleveland, OH 8.1%	Paterson, NJ 8.2%	Paterson, NJ 7.8%

Panel B: Change in Educational Attainment (Percentage Points)

	1970s	1980s	1990s	2000s	Total
Metropolitan Areas	6.0	4.3	4.4	2.7	17.4
Highest	Austin, TX 10.3	Bridgeport, CT 8.3	San Jose, CA 7.6	Madison, WI 5.3	Boston, MA 26.4
Lowest	El Paso, TX 2.6	El Paso, TX 1.2	Fresno, CA 0.6	Mobile, AL 0.6	El Paso, TX 6.4
Cities	5.8	4.00	3.3	2.7	15.8
Highest	Arlington, VA 12.9	Seattle, WA 9.8	San Francisco, CA 10	Arlington, VA 7	Arlington, VA 37.6
Lowest	Paterson, NJ 1.6	Santa Ana, CA -1.4	Hartford, CT -2.0	Rockford, IL -2.0	Santa Ana, CA 3.0

In 1970, 11.9% of the residents in metropolitan areas and 11.4% of the residents in central cities had at least a bachelor's degree. The metropolitan area with the highest level of educational attainment was Washington, DC, with 22.1% of residents having at least a bachelor's degree. Arlington, VA, with 29.6% of residents having at least a bachelor's degree, was the city with the highest level of educational attainment. Youngstown is the metropolitan area with the lowest educational attainment with 7.1% of its population having at least a bachelor's degree, while Newark, with only 4.3% of its residents with at least a bachelor's degree, has the lowest level among cities.

During the 1970s, the average metropolitan area experienced a 6.0 percentage point increase in the percentage of residents with at least a bachelor's degree while the average city had an increase of 5.8 percentage points. The educational gap between metropolitan areas and central cities increased slightly from 0.5 to 0.7 points. Among metropolitan areas, the largest increase in educational attainment was for Austin, TX, where the percentage of residents with at least a bachelor's degree increased by 10.3 percentage points. Austin was the only metropolitan area in which the increase exceeded 10 percentage points. Among cities, the largest increase was for Arlington, VA, which had an increase of 12.9 percentage points. There were five cities (Arlington, Seattle, San Francisco, Minneapolis, and Boston) with an increase in educational attainment that exceeded 10 points. The smallest increase in educational attainment for metropolitan areas was for El Paso, where the percentage of residents with at least a college degree increased by only 2.9 percentage points. Paterson, NJ, with a 1.6 percentage point increase, had the smallest increase among cities. In 1980, Washington, DC, remained the metropolitan area with the highest educational attainment, with 31.2% of its residents having at least a bachelor's degree, while Arlington also remained the city with the highest educational attainment at 42.5%. Youngstown remained the metropolitan area with the lowest educational attainment at 10.7% and Paterson remained the city with the lowest level at 6.2%.

During the 1980s, the educational attainment in the average metropolitan area increased by 4.3 percentage points to 22.2%, while for the average city, it increased by 4.0 points to 21.1%. The educational attainment gap increased by 0.3 percentage points to 1.1 percentage points. The Bridgeport, CT, metropolitan area had the largest increase among metropolitan areas with an 8.3 percentage point increase, while Seattle, with a 9.8 percentage point increase, had the largest increase among cities. El Paso, with a 1.2 percentage point increase, once again had the smallest increase among metropolitan areas, while Santa Ana, with a 1.4 percentage point decrease, had the largest decrease in educational attainment in the 1980s. There were two cities (Santa Ana and Miami) in which educational attainment decreased during the 1980s. In 1990, Washington, DC, remained the metropolitan area with the highest level of educational attainment with 37.5% of its population having at least a bachelor's degree, while Arlington remained the city with the highest level at 52.3%. Flint replaced Youngstown as the metropolitan area with the lowest educational attainment at 12.8%, while Cleveland replaced Paterson as the city with the lowest level at 8.1%.

During the 1990s, the educational attainment in the average metropolitan area increased by 4.4 percentage points to 26.5% while the educational attainment in the average city increased by 3.3 percentage points to 24.5%. The educational attainment gap increased by 1.0 points to 2.1 percentage points. This was the largest increase in the educational attainment gap in any of the decades. San Jose had the largest increase among metropolitan areas with an increase of 7.6 percentage points, while San Francisco had the largest increase among cities with a 10.0 percentage point increase. Fresno had the smallest increase among metropolitan areas with an increase of only 0.6 points, while Hartford had the largest decrease among cities with a −2.0 decrease. In the 1990s, there were six cities (Hartford, Santa Ana, Paterson, Riverside, Fresno, and Bridgeport) in which educational attainment decreased. In 2000, Washington remained the metropolitan area with the highest educational attainment with 42.5% of its residents having at least a bachelor's degree and Arlington remained the city with the highest level at 60.2%. Flint remained the metropolitan area with the lowest educational attainment at 16.2%, while Paterson, NJ, replaced Cleveland as the city with the lowest level at 8.2%.

During the 2000s, the educational attainment in the average metropolitan area increased by 2.7 percentage points to 29.2%. The educational attainment in the average city also increased by 2.7 points to 27.1%. The educational attainment gap held steady at 2.1 percentage points. The largest increase in educational attainment among metropolitan areas was for Madison, WI, with an increase of 5.3 percentage points. The largest increase for cities was for Arlington with an increase of 7.0 percentage points. The smallest increase for metropolitan areas was for Mobile, AL, which had an increase of 0.6 percentage points, while Rockford, IL, had the largest decrease among cities with a decrease of 2.0 points. There were seven cities (Rockford, Dallas, Jackson (MS), Paterson, Houston, Baton Rouge, and Memphis) where educational attainment decreased in the 2000s. In 2010, Washington and Arlington retained their positions as the metropolitan area and cities with the highest educational attainment. In the Washington metropolitan area, 46.1% of its residents aged 25 and over had at least a bachelor's degree, while in Arlington, the figure was 67.2%. Youngstown was the metropolitan area with the lowest educational attainment at 17.4%, while Paterson remained the city with the lowest educational attainment at 7.8%.

Table 3.8 contains the metropolitan areas and cities with the largest and smallest changes in educational attainment from 1970 to 2010. The metropolitan areas with the largest absolute increases in educational attainment were Boston (26.4 points), San Francisco (25.6 points), and Bridgeport, CT (24.9 points). There were 21 metropolitan areas in which the percentage of residents 25 and over increased by more than 20 percentage points. In terms of percentage change, the largest increases in educational attainment were for Worcester, MA (256%), Providence, RI (234%), and Baltimore (226%). Among cities the largest absolute increases in educational attainment were for Arlington (37.6 points), Seattle (37.1 points), and

TABLE 3.8 Total Change in Educational Attainment, 1970–2010

Largest Increase	Percentage Points		Percentage Change	
	MSA	Change	MSA	% Change
	Boston, MA	26.4	Worcester, MA	256%
	San Francisco, CA	25.6	Providence, RI	234%
	Bridgeport, CT	24.9	Baltimore, MD	226%
Smallest Increase	Percentage Points		Percentage Change	
	MSA	Change	MSA	% Change
	El Paso, TX	6.4	El Paso, TX	56%
	Riverside, CA	8.3	Riverside, CA	85%
	Corpus Christi, TX	8.7	Fresno, CA	86%
Largest Increase	Percentage Points		Percentage Change	
	City	Change	City	% Change
	Arlington, VA	37.6	Jersey City, NJ	479%
	Seattle, WA	37.1	St. Louis, MO	325%
	San Francisco, CA	33.7	Boston, MA	304%
Smallest Increase	Percentage Points		Percentage Change	
	City	Change	City	% Change
	Santa Ana, CA	3	Riverside, CA	30%
	Paterson, NJ	3.2	Santa Ana, CA	38%
	Riverside, CA	4.9	Jackson, MS	59%
	Youngstown, OH	4.9		

San Francisco (33.7 points). It is worth noting that, in addition to having the largest increase in educational attainment, Arlington had the highest initial education attainment level among cities in 1970. In terms of percentage change, the largest increases were for Jersey City (479%), St. Louis (325%), and Boston (304%). The metropolitan areas with the smallest absolute increases in educational attainment were El Paso (6.4 points), Riverside (8.3 points), and Corpus Christi (8.4 points). In percentage terms, the smallest increases were for El Paso (56%), Riverside (85%), and Fresno (86%). Among cities, the smallest increases in absolute educational attainment were for Santa Ana (3.0 points), Paterson (3.2 points), and Riverside and Youngstown (4.9 points). In percentage terms, the smallest increases were for Riverside (30%), Santa Ana (38%), and Jackson, MS (59%).

Tables 3.7 and 3.8 identified the metropolitan areas and central cities with the highest and lowest levels of educational attainment, as well as the areas that experienced the largest and smallest changes in educational attainment. Table 3.9 focuses

TABLE 3.9 Relative Educational Attainment

Panel A: Central City Educational Attainment as a Percentage of Metropolitan Educational Attainment

	1970		1980		1990		2000		2010	
	City	*%*	*City*	*%*	*City*	*%*	*City*	*%*	*City*	*%*
Highest	Greensboro, NC	181%	Greensboro, NC	160%	Greensboro, NC	157%	Greensboro, NC	145%	Seattle, WA	148%
	Riverside, CA	168%	Charlotte, NC	147%	Baton Rouge, LA	140%	Seattle, WA	144%	Arlington, VA	146%
	Baton Rouge, LA	155%	Virginia Beach, VA	145%	Seattle, WA	140%	Arlington, VA	142%	Greensboro, NC	144%
	MSAs	*%*	*City*	*%*	*MSAs*	*%*	*City*	*%*	*MSAs*	*%*
Lowest	Bridgeport, CT	32%	Paterson, NJ	32%	Newark, NJ	33%	Paterson, NJ	33%	Paterson, NJ	23%
	Newark, NJ	35%	Newark, NJ	33%	Paterson, NJ	34%	Newark, NJ	34%	Bridgeport, CT	33%
	Paterson, NJ	37%	Bridgeport, CT	34%	Bridgeport, CT	36%	Bridgeport, CT	36%	Newark, NJ	34%

Panel B: Change in Central City Educational Attainment as a Percentage of Metropolitan Educational Attainment

	1970s		1980s		1990s		2000s		Total	
	Cities	*Change*	*Cities*	*Change*	*Cities*	*Change*	*Cities*	*Change*	*Cities*	*Change*
Highest	Minneapolis, MN	21.8	Jersey City NJ	23.0	Tampa, FL	9.0	Fort Wayne, IN	14.0	Jersey City, NJ	49.7
	Boston, MA	18.5	Seattle, WA	12.7	San Francisco, CA	7.3	New Orleans, LA	12.1	Seattle, WA	37.0
	Seattle, WA	16.3	Tampa, FL	12.6	Jersey City NJ	6.8	Miami, FL	11.4	Boston, MA	29.9
	MSAs	*Change*	*City*	*Change*	*MSAs*	*Change*	*City*	*Change*	*Change*	*City*
Lowest	Riverside, CA	-24.5	Santa Ana, CA	-17.1	Hartford, CT	-13.7	Rockford, IL	-15.4	Riverside, CA	-50.1
	Greensboro, NC	-21.1	Virginia Beach, VA	-15.1	Riverside, CA	-13.2	Baton Rouge, LA	-9.3	Jackson, MS	-41.3
	Evansville, IN	15.1	Miami, FL	-14.1	Jackson, MS	-12.0	Tulsa, OK	-9.0	Greensboro, NC	-37.2

on the relative levels of educational attainment between the central cities and their metropolitan areas. The table identifies the cities with the highest and lowest educational attainment relative to their metropolitan areas for each decade, as well as the cities that experienced the biggest increases and decreases in relative educational attainment in each decade and over the entire 1970–2010 period.

In 1970, Greensboro, NC, had the highest level of relative educational attainment with the city's level being 81% higher than the metropolitan area. The next highest levels were for Riverside (68% higher) and Baton Rouge (55% higher). The lowest levels were for Bridgeport (68% lower), Newark (65% lower), and Paterson (63% lower).

During the 1970s, the largest increases in relative educational attainment were for Minneapolis (22 points), Boston (19 points), and Seattle (16 points), while the largest decreases were for Riverside (25 points), Greensboro (21 points), and Evansville (15 points). By 1980, Greensboro, despite its decrease in relative educational attainment in the 1970s, still had the highest level of relative educational attainment with the city's educational attainment being 60% higher than the metropolitan area's attainment. The next highest levels were for Charlotte (47% higher) and Virginia Beach (45% higher). The lowest levels of relative educational attainment in 1980 were for Paterson (68% lower), Newark (67% lower), and Bridgeport (66% lower).

During the 1980s, the largest increases in relative educational attainment were for Jersey City (23 points), Seattle (13 points), and Tampa (13 points). Thus, Seattle was among the cities with the largest increases in both the 1970s and the 1980s. The largest decreases were for Santa Ana (17 points), Virginia Beach (15 points), and Miami (14 points). In 1990, Greensboro remained the city with the highest level of relative educational attainment with the city's level exceeding the metropolitan area's level by 56%. The next highest levels were for Baton Rouge (40% higher) and Seattle (40% higher). Newark had the lowest level of relative educational attainment with the city's educational attainment being 67% below the metropolitan area's educational attainment. The next lowest levels were for Paterson (66% lower) and Bridgeport (64% lower).

During the 1990s, the largest increases in relative educational attainment were for Tampa (9 points), San Francisco (7 points), and Jersey City (7 points). Both Tampa and Jersey City were also among the cities with the largest increases in the 1980s. The largest decreases in relative educational attainment were for Hartford (14 points), Riverside (13 points), and Jackson, MS (12 points). Riverside was also among the cities with the largest decreases in the 1970s. In 2000, Greensboro was, once again, the city with the highest level of relative educational attainment with the city's level 45% higher than the metropolitan area's level. The next highest levels were for Seattle (44% higher) and Arlington (42% higher). Paterson had the lowest level of relative educational attainment with the city's level being 73% lower than the level for the metropolitan area. The next lowest levels were for Newark (70% lower) and Bridgeport (69% lower).

During the 2000s, the largest increases in relative educational attainment were for New Orleans (12 points), Miami (11 points), and Sacramento (10 points). The large increase in New Orleans, coupled with the massive population losses in the 2000s due to Hurricane Katrina, suggests that the outmigration of residents was heavily concentrated among less-educated residents. The largest decreases in relative educational attainment were for Rockford, IL (15 points), Baton Rouge (9 points), and Tulsa, OK (9 points). In 2010, Seattle had surpassed Greensboro as the city with the highest level of relative educational attainment with the city's attainment exceeding the metropolitan area's attainment by 48%. Arlington also passed Greensboro to become the city with the second highest level of relative educational attainment with a level that exceeded the metropolitan area's attainment by 46%. Greensboro now had the third highest level with the city's educational attainment being 44% higher than the metropolitan area's attainment. Paterson remained the city with the lowest level of relative educational attainment with a level that was 77% lower than the educational attainment of the metropolitan area. The next lowest levels were for Bridgeport (67% lower) and Newark (66% lower).

The final column of Panel B of Table 3.9 identifies the cities with the largest increases and decreases in relative educational attainment between 1970 and 2010. Jersey City had the largest increase with a 50-percentage point increase. The next largest increases were for Seattle (37 points) and Boston (30 points). Riverside had the largest decrease in relative educational attainment with a decrease of 50 percentage points. The next largest decreases were for Jackson, MS (41 points) and Greensboro (37 points).

The final table regarding educational attainment, Table 3.10 shows the metropolitan areas and cities with the highest and lowest percentage changes in the absolute number of residents aged 25 and above with at least a bachelor's degree. In the 1970s, the metropolitan areas with the highest growth in the number of college-educated residents were Austin (165%), Houston (146%), and Phoenix (144%), while the metropolitan areas with the lowest levels of growth were Syracuse (48%), Cleveland (49%), and Buffalo (52%). The cities with the highest growth in the number of college graduates were Virginia Beach (192%), Austin (130%), and Charlotte (128%), while the cities with the lowest growth were Detroit (9%), Flint (13%), and Cleveland (19%).

In the 1980s, the metropolitan areas with the highest growth in the number of residents with college degrees were Riverside (95%), Atlanta (92%), and Austin (85%). The lowest growth among metropolitan areas was recorded for Youngstown (24%), Shreveport (26%), and Flint (27%); among cities, the highest growth was for Jersey City (107%), Fresno (89%), and Virginia Beach (88%), while the lowest growth was for Youngstown (0.8%), Gary (1.6%), and Miami (2.6%).

In the 1990s, Austin once again had the highest growth among metropolitan areas with an increase of 81%. The next largest increases were for Charlotte (80%) and Atlanta (73%). Austin had the highest growth rate in both the 1970s and 1990s and the third-highest growth in the 1980s. Atlanta had the second-highest growth

TABLE 3.10 Percentage Change in Number of College-educated Residents

	1970s		1980s		1990s		2000s		Total	
Highest	*Metro Area*	*% Growth*	*Metro Area*	*% Growth*	*Metro Area*	*% Growth*	*Metro Area*	*% Growth*	*Metro Area*	*% Growth*
	Austin, TX	165%	Riverside, CA	95%	Austin, TX	81%	Riverside, CA	41%	Austin, TX	1060%
	Houston, TX	146%	Atlanta, GA	92%	Charlotte, NC	80%	Phoenix, AZ	36%	Phoenix, AZ	926%
	Phoenix, AZ	144%	Austin, TX	85%	Atlanta, GA	73%	Austin, TX	31%	Atlanta, GA	852%
Lowest	*Metro Area*	*% Growth*	*Metro Area*	*% Growth*	*Metro Area*	*% Growth*	*Metro Area*	*% Growth*	*Metro Area*	*% Growth*
	Syracuse, NY	48%	Youngstown, OH	24%	Lubbock, TX	15%	New Orleans, LA	–10%	Syracuse, NY	167%
	Cleveland, OH	49%	Shreveport, LA	26%	Syracuse, NY	18%	Youngstown, OH	5%	Dayton, OH	170%
	Buffalo, NY	52%	Flint, MI	27%	Dayton, OH	18%	Dayton, OH	5%	Youngstown, OH	170%

(Continued)

TABLE 3.10 (Continued)

Highest	1970s City	% Growth	1980s City	% Growth	1990s City	% Growth	2000s City	% Growth	Total City	% Growth
	Virginia Beach, VA	192%	Jersey City, NJ	107%	Charlotte, NC	76%	Riverside, CA	42%	Charlotte, NC	769%
	Austin, TX	130%	Fresno, CA	89%	Austin, TX	69%	Sacramento, CA	38%	Virginia Beach, VA	713%
	Charlotte, NC	128%	Virginia Beach, VA	88%	Portland, OR	54%	Newark, NJ	34%	Austin, TX	691%

Lowest	1970s City	% Growth	1980s City	% Growth	1990s City	% Growth	2000s City	% Growth	Total City	% Growth
	Detroit, MI	9%	Youngstown, OH	1%	Hartford, CT	−25%	New Orleans, LA	−40%	Youngstown, OH	8%
	Flint, MI	13%	Gary, IN	2%	Jackson, MS	−8%	Youngstown, OH	−11%	Detroit, MI	9%
	Cleveland, OH	19%	Miami, FL	3%	Syracuse, NY	−7%	Rockford, IL	−9%	Gary, IN	29%

in the 1980s and the third-highest growth in the 1990s. The lowest growth rates among metropolitan areas were for Lubbock (15%), Syracuse (18%), and Dayton (18%). Among cities, Charlotte had the highest growth rate at 76% and was followed by Austin (69%) and Portland (54%). In the 1990s, for the first time, there were cities in which the number of residents with at least a bachelor's degree declined. The largest decreases were for Hartford, which had a decrease of 25%, Jackson, MS (8% decrease), and Syracuse (7% decrease). In all, there were nine cities in which the number of residents with at least a bachelor's degree decreased during the 1990s.

In the 2000s, Riverside was the metropolitan area with the largest increase in the number of college graduates with an increase of 41% followed by Phoenix (36%) and Austin (31%). Thus, Austin was among the metropolitan areas with the largest increases in the number of residents with at least a bachelor's degree in every decade included in this study. The New Orleans metropolitan area experienced a 10% decrease in the number of college graduates in the 2000s, but, as has been mentioned before, this is primarily due to the impact of Hurricane Katrina. New Orleans was the only metropolitan area with a decrease in the number of college graduates and the areas with the smallest increases were Youngstown (5%) and Dayton (5%). Dayton had the third-smallest increase in both the 1990s and 2000s. Among cities, Riverside had the largest increase with an increase of 42%, while the next-largest increases were for Sacramento (38%) and Newark (34%). In the 2000s, there were, once again, several cities that experienced a decline in the number of residents with at least a bachelor's degree. New Orleans had the largest decline at 40% while the next largest decreases were for Youngstown (11%) and Rockford (9%). In all, there were 16 cities in which the number of residents aged 25 and over with at least a bachelor's degree decreased in the 2000s.

The final column of the table identifies the metropolitan areas and cities with the largest and smallest cumulative changes from 1970 to 2010. Given the results from the individual decades, it is not surprising that Austin was the metropolitan area with the largest increase in the number of residents with at least a bachelor's degree. Austin had a 1060% increase between 1970 and 2010 and the next largest increases were for Phoenix (926%) and Atlanta (852%). The smallest increases among metropolitan areas were for Syracuse (167%), Dayton (170%), and Youngstown (170%). For cities, the largest increases over the entire 1970–2010 period were for Charlotte (769%), Virginia Beach (713%), and Austin (691%), while the smallest increases were for Youngstown (8%), Detroit (9%), and Gary (29%).

Metropolitan Trends in Occupation

The final set of tables in this chapter summarizes the trends with respect to the number of residents employed in professional and executive occupations in the metropolitan areas and cities in the sample. The tables in this section follow the same general patterns as the tables regarding educational attainment. First, the metropolitan and central city averages will be compared over time. Second, the areas with the highest and lowest percentages of employed residents in professional and executive occupations are identified as well as those that experienced the largest and smallest increases in the percentage for each decade and over the entire 1970–2010 period. Next, the cities with the highest and lowest percentage relative to their metropolitan areas are identified and then, finally, the areas with the highest and smallest increases in the absolute numbers of residents employed in these occupations are identified.

Table 3.11 contains the average percentages of residents employed in professional and executive occupations over time for the metropolitan areas and cities in the sample as well as the areas with the highest and lowest values. In 1970, 24.7% of employed metropolitan residents and 23.5% of employed residents of cities were in professional and executive occupations. For metropolitan areas, the highest concentrations were in Los Angeles (35.9%), Washington (35.1%), and San Jose (32.9%), while the lowest values were for Flint (17.5%), Youngstown (18.4%), and Greensboro (18.9%). Among cities, the highest values were for Arlington (40.9%), Virginia Beach (40.9%), and Albuquerque (35.3%), while the lowest values were for Newark (12.1%), Cleveland (13.2%), and Paterson (13.3%).

During the 1970s, the percentage of employed residents in professional and executive occupations for metropolitan areas increased by 2.4 percentage points to 27.1%, while the average for cities increased by 2.7 percentage points to 26.2%. Thus, the occupational gap decreased from 1.2 points to 0.9 points. The largest increase among metropolitan areas was for Baltimore which had an increase of 4.5 percentage points. The next largest increases were for Worcester (4.4 points) and Dayton (4.2 points). There were three metropolitan areas in which the percentage decreased during the 1970s. These areas were Albuquerque (−0.8), El Paso (−0.6), and Corpus Christi (−0.3). Among cities the largest increases were for Washington, DC (10.0 points), Minneapolis (7.6 points), and Boston (7.3 points). There were eight cities in which the percentage of workers employed in professional and executive occupations decreased in the 1970s. The largest decreases were for Riverside (−1.6), Albuquerque (−1.5), and Shreveport (−0.9). In 1980, Washington, DC, became the metropolitan area with the highest percentage of workers in professional and executive occupations at 39.1%, while Flint remained the metropolitan area with the lowest percentage at 20.2%. Arlington remained the city with the highest percentage of workers employed in professional and executive occupations at 47.8%. Virginia Beach also had 47.8% of its workers employed in these occupations while Madison, WI, had the next highest value at 38.3%. Paterson became the city with the lowest value at 13.2%, while Newark (14.2%) and Gary (14.9%) had the next lowest values.

During the 1980s, the percentage of workers employed in professional and executive occupations for the average metropolitan area increased by 4.5 percentage points to 31.6%, while the percentage for the average city increased by 4.2 percentage points to 30.4%. Thus, the occupational gap increased by 0.4 points to 1.2

TABLE 3.11 Percentage of Workers Employed in Professional and Executive Occupations

Panel A: % of Employed Residents Working in Professional and Executive Occupations

	1970	1980	1990	2000	2010
Metropolitan Areas	24.7%	27.1%	31.6%	35.0%	35.4%
Highest	Los Angeles, CA	Washington, DC	Washington, DC	Washington, DC	Washington, DC
	35.9%	39.1%	44.6%	49.1%	50.0%
Lowest	Flint, MI	Flint, MI	Youngstown, OH	Youngstown, OH	Rockford, IL
	17.5%	20.2%	24.1%	26.1%	26.5%
Cities	23.5%	26.2%	30.4%	32.5%	34.4%
Highest	Arlington, VA	Arlington, VA	Virginia Beach, VA	Arlington, VA	Washington, DC
	40.9%	47.8%	55.0%	61.3%	56.1%
Lowest	Newark, NJ	Paterson, NJ	Paterson, NJ	Santa Ana, CA	Santa Ana, CA
	12.1%	13.2%	16.0%	16.5%	15.9%

Panel B: Change in Percentage of Workers Employed in Professional and Executive Occupations (Percentage points)

	1970s	1980s	1990s	2000s	Total
Metropolitan Areas	2.39	4.54	3.37	0.37	10.66
Highest	Baltimore	Worcester, MA	San Jose, CA	Pittsburgh, PA	Baltimore, MD
	4.5	7.2	7.3	2.6	18.1
Lowest	Albuquerque, NM	Tucson, AZ	Shreveport, LA	Fresno, CA	Riverside, CA
	-0.8	2.5	0.3	-2.7	3.0
Cities	2.71	4.17	2.14	1.86	10.89
Highest	Washington, DC	Seattle, WA	San Francisco, CA	Washington, DC	Washington, DC
	10.0	8.0	9.9	5.0	28.8
Lowest	Riverside, CA	Santa Ana, CA	Hartford, CT	Dallas, TX	Santa Ana, CA
	-1.6	-1.4	-2.6	-4.5	-3.5

percentage points in 1990. The metropolitan areas with the largest increases in the percentage of workers in professional and executive occupations were Worcester (7.2 points), Boston (6.9 points), and Baltimore (6.8 points), while the metropolitan areas with the smallest increases were Tucson (2.5 points) and four cities (Miami, Toledo, Riverside, and Corpus Christi) with increases of 2.7 percentage points. Among cities, the largest increases during the 1980s were for Seattle (8.0 points), Jersey City (7.8 points), and Virginia Beach and Arlington (7.2 points). Santa Ana, with a decline of 4.0 percentage points, was the only city in which the percentage decreased, while the smallest increases were for Miami (0.1 points) and Anaheim (0.5 points). In 1990, Washington remained the metropolitan area with the highest percentage of workers employed in professional and executive occupations at 44.6%, while Youngstown became the metropolitan area with the lowest percentage at 24.1%. Virginia Beach and Arlington remained the cities with the highest percentage at 55.0%, while Paterson remained the city with the lowest percentage at 16.0%.

During the 1990s, the percentage of workers employed in professional and executive occupations in the average metropolitan area increased by 3.4 percentage points to 35.0%, while the percentage for the average city increased by 2.1 percentage points to 32.5%. The occupational gap increased by 1.2 points to 2.5 percentage points. San Jose, with an increase of 7.3 percentage points, had the largest increase among metropolitan areas and was followed by Charlotte (6.4 points) and Madison (6.0 points). Shreveport had the smallest increase at 0.3 points and was followed by Dayton (1.1 points) and Salt Lake City (1.2 points). San Francisco had the largest increase among cities with an increase of 9.9 points, while Atlanta (8.2 points) and Los Angeles (7.7 points) had the next largest increases. There were 13 cities in which the percentage of workers employed in professional and executive occupations decreased during the 1990s. The largest decreases were for Hartford (−2.6 points), Youngstown (−2.2 points), and Toledo (−0.9 points). In 2000, Washington remained the metropolitan area with the highest percentage at 49.1%, while Youngstown had the lowest percentage at 26.1%. Among cities, Arlington continued to have the highest percentage at 61.3%, while Santa Ana had the lowest percentage at 16.5%.

During the 2000s, the percentage of workers employed in professional and executive occupations in the average metropolitan area increased by 0.4 percentage points to 35.4% while the percentage in the average city increased by 1.9 percentage points to 34.4%. The occupational gap decreased by 1.5 points to 1.0 percentage points. Among metropolitan areas, Pittsburgh and Columbus, OH, had the largest increases in the percentage of workers employed in professional and executive occupations with a 2.6 percentage points increase. The next largest increase among metropolitan areas was for Omaha, which had a 2.5 percentage points increase. There were 24 metropolitan areas in which the percentage of workers employed in professional and executive occupations decreased during the 2000s. The largest decreases were for Fresno (2.7 points), Rockford (2.6 points), and Dallas

(1.9 points). Among cities, the largest increases were for Washington (5.0 points), Tacoma (4.3 points), and St. Louis (4.2 points). There were 27 cities in the sample for which the percentage of workers employed in professional and executive occupations decreased during the 2000s.[2] The largest decreases among cities were for Dallas (−4.5 points), Fresno (−3.7 points), and Houston (−3.3 points). In 2010, Washington remained the metropolitan area with the highest percentage of workers employed in professional and executive occupations at 50.0%. Rockford had the lowest percentage among metropolitan areas at 26.5%. Among the cities with occupational data available for 2010, Washington had the highest percentage of workers employed in professional and executive occupations at 56.1%, while Santa Ana had the lowest percentage at 15.9%.

The final column of Panel B of Table 3.11 contains the metropolitan areas and cities with the largest and smallest increases in the percentage of workers employed in professional and executive occupations from 1970 to 2010. The metropolitan areas with the largest increases were Baltimore (18.1 points), San Francisco (16.3 points), and Charlotte (16.2 points). The smallest increases were for Riverside (3.0 points), El Paso (3.5), and Corpus Christi (3.6 points). Among the cities with occupational data available for 2010, the largest increases were for Washington (28.8 points), San Francisco (25.9 points), and Seattle (24.1 points). There were two cities (Santa Ana and Riverside) in which the percentage of workers employed in professional and executive occupations decreased between 1970 and 2010. Santa Ana had the largest decrease at 3.5 points, while the decline in Riverside was 1.4 points.

Table 3.12 focuses on the percentage of workers in each city that are employed in professional and executive occupations relative to the percentage in the metropolitan area. The goal is to identify the cities in which professional and executive workers are most overrepresented or underrepresented. For each city, the city percentage of workers in professional and executive occupations is divided by the percentage for the metropolitan area containing the city. For each year, the cities with the highest and lowest ratios are calculated. In addition, the cities with the largest and smallest changes during each decade and over the entire 1970–2010 period are calculated.

In 1970, the cities with the highest ratios were Greensboro (143%), Charlotte (131%), and Riverside (128%), while the cities with the lowest ratios were Newark (46%), Paterson (51%), and Bridgeport (55%). In total, there were 42 cities in which the city's share of workers in professional and executive occupations exceeded the percentage for the metropolitan area.

During the 1970s, the largest increase in the occupation ratio was for Washington, DC, and Minneapolis. Most cities had a 17.6-percentage point increase. The next largest increase was for Los Angeles (15.2 points). The largest decreases in the ratio were for Warren, MI (−14.1 points), Greensboro (−9.3 points), and Flint (−9.2 points). In 1980, Greensboro remained the city with the highest occupation ratio at 134%. The next highest values were for Charlotte (126%) and Virginia Beach (125%). The lowest values in 1980 were still for Paterson (45%), Newark (48%), and Bridgeport (50%). In 1980, there were 49 cities in which the percentage

TABLE 3.12 Ratio of City Percentage of Workers in Professional and Executive Occupations to Metropolitan Percentage

Panel A: Central City Percentage Professional and Executive to Metropolitan % Professional and Executive

	1970 City	%	1980 City	%	1990 City	%	2000 City	%	2010 City	%
Highest	Greensboro, NC	143%	Greensboro, NC	134%	Greensboro, NC	126%	Arlington, VA	126%	Seattle, WA	127%
	Charlotte, NC	131%	Charlotte, NC	126%	Arlington, VA	123%	Seattle, WA	123%	Salt Lake City, UT	118%
	Riverside, CA	128%	Virginia Beach, VA	125%	Seattle, WA	119%	Greensboro, NC	119%	San Francisco, CA	117%
Lowest MSAs	City	%	City	%	City	%	City	%	City	%
	Newark, NJ	46%	Paterson, NJ	45%	Paterson, NJ	45%	Paterson, NJ	46%	Santa Ana, CA	47%
	Paterson, NJ	51%	Newark, NJ	48%	Newark, NJ	48%	Santa Ana, CA	48%	Detroit, MI	64%
	Bridgeport, CT	55%	Bridgeport, CT	50%	Santa Ana, CA	50%	Newark, NJ	52%	Cleveland, OH	69%

Panel B: Change in Ratio Central City Professional and Executive Employment Percentage to Metropolitan Professional and Executive Employment Percentage

	1970s Cities	Change	1980s Cities	Change	1990s Cities	Change	2000s Cities	Change	Total Cities	Change
Highest	Washington, DC	17.6	Jersey City, NJ	12.5	Los Angeles, CA	14.1	Fort Lauderdale, FL	11.1	Washington, DC	34.4
	Minneapolis, MN	17.6	Gary, IN	11.2	Atlanta, GA	9.1	Yonkers, NY	9.9	Jersey City, NJ	32.1
	Los Angeles, CA	15.2	Seattle, WA	8.0	San Francisco, CA	9.0	Fort Wayne, IN	9.5	Seattle, WA	26.5
Lowest MSAs	City	Change	City	Change	City	Change	City	Change	City	Change
	Riverside, CA	-24.5	Santa Ana, CA	-17.1	Hartford, CT	-17.1	Rockford, IL	-13.7	Riverside, CA	-50.1
	Greensboro, NC	-21.1	Virginia Beach, VA	-15.1	Riverside, CA	-15.1	Baton Rouge, LA	-13.2	Jackson, MS	-41.3
	Evansville, IN	-15.1	Miami, FL	-14.1	Jackson, MS	-14.1	Tulsa, OK	-12.0	Greensboro, NC	-37.2

of workers employed in professional and executive occupations exceeded that of their metropolitan areas. This was an increase over the 42 cities in 1970.

During the 1980s, the largest increases in the occupation ratio were for Jersey City (12.5 points), Gary, IN (11.2 points), and Seattle (8.0 points). The largest decreases were for Santa Ana (19.8 points), Virginia Beach (11.9 points), and Jackson, MS (9.9 points). In 1990, Greensboro remained the city with the highest ratio at 126%. The next highest values were for Arlington, VA (123%) and Seattle (119%). The lowest values in 1990, once again, were for Paterson (46%) and Newark (48%). Santa Ana replaced Bridgeport as the city with the third-lowest value at 52%. In 1990, there were 44 cities in which the central city percentage of workers in professional and executive occupations exceeded the percentage for the metropolitan area. This was a decrease from the 49 cities in 1980.

During the 1990s, the largest increases in the occupation ratio were for Los Angeles (14.1 points), Atlanta (9.1 points), and San Francisco (9.0 points). The largest decreases were for Youngstown (−15.1 points), Hartford (−13.6 points), and Jackson, MS (−11.2 points). In 2000, both Arlington (125%) and Seattle (125%) surpassed Greensboro as the cities with the highest occupation ratios. Greensboro had the third-highest ratio at 118%. The lowest values in 2000 were for Paterson (43%), Santa Ana (47%), and Newark (49%). In 2000, there were 36 cities in which the percentage of workers in professional and executive occupations exceeded the percentage for their metropolitan areas. This was a decline from the 44 cities in 1990 and is the lowest value so far.

During the 2000s, the largest increases in the occupation ratio, for cities for which occupational data was available in 2010, were for Fort Lauderdale (11.1 points), Yonkers (9.9 points), and St. Louis (9.2 points), while the largest decreases were for Cincinnati (−8.2 points), Dallas (−8.1 points), and Omaha (−7.6 points). In 2010, the highest occupation ratios, among the cities for which occupation data was available, were for Seattle (127%), Salt Lake City (118%), and San Francisco (117%), while the lowest ratios were for Santa Ana (47%), Detroit (64%), and Cleveland (69%).

The final column of Panel B of Table 3.12 reveals the cities with the largest increases and decreases in their occupation ratio between 1970 and 2010. The largest increases were for Washington, DC (34.4 points), Jersey City (32.0 points), and Seattle (26.5 points), while the largest decreases were for Greensboro (−31.4 points), Santa Ana (−24.7 points), and Charlotte (−24.5 points).

The final table in this chapter, Table 3.13, identifies the metropolitan areas and cities with the largest and smallest percentage changes in the number of workers employed in professional and executive occupations for each decade and over the entire 1970 to 2010 period. In 1970s, the largest increases in the number of professional and executive workers among metropolitan areas were for Austin (111%), Phoenix (109%), and Tucson (107%). There were three metropolitan areas in which the number of professional and executive workers decreased. They were Buffalo (−3.9%), Youngstown (−4.7%), and Cleveland (−1.6%). Among cities, the largest increases in the 1970s were for Virginia Beach (154%), San Jose (105%), and

TABLE 3.13 Percentage Change in Number of Workers Employed in Professional and Executive Occupations

Highest	1970s		1980s		1990s		2000s		Total	
	Metro Area	% Growth	Metro Area	% Growth	Metro Area	% Growth	Metro Area	% Growth	Metro Area	% Growth
	Austin, TX	111%	Riverside, CA	97%	Austin, TX	70%	Riverside, CA	31%	Austin, TX	662%
	Phoenix, AZ	109%	Austin, TX	85%	Charlotte, NC	54%	Phoenix, AZ	29%	Phoenix, AZ	580%
	Tucson, AZ	107%	Atlanta, GA	75%	Atlanta, GA	54%	Charlotte, NC	22%	Atlanta, GA	454%
Lowest	1970s		1980s		1990s		2000s		Total	
	Metro Area	% Growth	Metro Area	% Growth	Metro Area	% Growth	Metro Area	% Growth	Metro Area	% Growth
	Buffalo, NY	19%	Youngstown, OH	14%	Syracuse, NY	5%	New Orleans, LA	-19%	Buffalo, NY	67%
	Youngstown, OH	25%	Shreveport, LA	16%	Los Angeles, CA	6%	San Jose, CA	-4%	Youngstown, OH	67%
	Cleveland, OH	26%	Corpus Christi, TX	19%	Dayton, OH	7%	Rockford, IL	-2%	New Orleans, LA	67%

(Continued)

TABLE 3.13 (Continued)

	1970s		1980s		1990s		2000s		Total	
	City	% Growth	City	% Growth	City	% Growth	City	% Growth	City	% Growth
Highest	Virginia Beach, VA	154%	Virginia Beach, VA	78%	Austin, TX	60%	Louisville, KY	151%	Virginia Beach, VA	511%
	San Jose, CA	105%	Fresno, CA	77%	Charlotte, NC	55%	Fort Wayne, IN	29%	Austin, TX	430%
	Houston, TX	90%	San Diego, CA	69%	Portland, OR	44%	Fort Lauderdale, FL	29%	Charlotte, NC	414%
	1970s		1980s		1990s		2000s		Total	
	City	% Growth	City	% Growth	City	% Growth	City	% Growth	City	% Growth
Lowest	Flint, MI	−13%	Youngstown, OH	−9%	Hartford, CT	−34%	New Orleans, LA	−49%	Detroit, MI	−30%
	Detroit, MI	−10%	Miami, FL	−4%	Youngstown, OH	−14%	Detroit, MI	−16%	New Orleans, LA	−27%
	Gary, IN	−7%	Detroit, MI	−3%	Syracuse, NY	−14%	Cincinnati, OH	−16%	Cleveland, OH	8%

Houston (90%), while the largest decreases among cities in the 1970s were for Flint (−13.4%), Detroit (−10.3%), and Gary (−7.5%).

During the 1980s, the largest increases among metropolitan areas were for Riverside (97%), Austin (85%), and Atlanta (75%). The smallest increases among metropolitan areas in the 1980s were for Youngstown (14%), Shreveport (16%), and Corpus Christi (19%). Among cities, the largest increases were for Virginia Beach (78%), Fresno (77%), and San Diego (69%). There were three cities in which the number of workers employed in professional and executive occupations decreased in the 1980s. They were Youngstown (−8.5%), Miami (−4.2%), and Detroit (−3.4%).

During the 1990s, the largest increases among metropolitan areas were for Austin (70%), Charlotte (54%), and Atlanta (54%), while the smallest increases were for Syracuse (4.9%), Los Angeles (6.4%), and Dayton (6.7%). Among cities, the largest increases during the 1990s were for Austin (60%), Charlotte (55%), and Portland (44%). The largest decreases among the cites were for Hartford (−33.6%), Youngstown (−14.3%), and Syracuse (−13.5%).

During the 2000s, the largest increases among metropolitan areas were for Riverside (31%), Phoenix (29%), and Charlotte (22%), while the biggest decreases were for New Orleans (−19.4%), San Jose (−4.1%), and Rockford (−1.7%). For cities, the largest increases, among cities with occupational data available in 2010, were for Fort Lauderdale (29.1%), Riverside (28.1%), and Miami (27.5%), while the largest decreases were for New Orleans (−48.9%), Detroit (−16.1%), and Cincinnati (−16.0%).

Finally, the last column of Table 3.13 contains the metropolitan areas and cities with the largest and smallest increases in the number of workers employed in professional and executive occupations between 1970 and 2010. Among metropolitan areas, the largest increases were for Austin (663%), Phoenix (580%), and Atlanta (454%), while the smallest increases were for Buffalo (66.7%), Youngstown (66.7%), and New Orleans (67.1%). Among cities, the largest increases, among cities with occupation data available in 2010, were for Virginia Beach (511%), Austin (430%), and Charlotte (414%). There were two cities (New Orleans and Detroit) for which the number of workers employed in professional and executive occupations decreased between 1970 and 2010. New Orleans experienced a decrease of 29.6%, while Detroit had a decrease of 27.2%.

Conclusion

This chapter has analyzed central city and metropolitan trends in population, employment, income, education, and occupation for 1970–2010. The goal was to establish the context within which the gentrification trends identified in the chapters that follow were taking place.

The primary takeaway from the analysis in this chapter is that the gentrification trends that are identified in the next three chapters take place at a time when the primary pattern for all the variables studied was one of suburbanization. For every variable, the overall metropolitan growth rates in every decade were higher than the central city growth rates. This, of course, means that suburbs were growing faster than central cities. While, in some cases, the rate of suburbanization slowed between 1970 and 2010, the gentrification activity identified in the next three chapters was not the result of a large-scale "return to the city" movement in which there was a reversal of the long-term suburbanization of the metropolitan population in the United States. Instead, the gentrification that occurred was the result of certain neighborhoods experiencing high levels of growth in income, educational attainment, and workers employed in professional and executive occupations despite the overall trend of higher-income households, college graduates, and professional and executive workers becoming more concentrated in the suburbs.

More specifically, the primary results in this chapter are as follows:

- The central city share of the population of the metropolitan areas included in this study fell by 4.0 percentage points between 1970 and 2010.
- The central city share of total employment decreased by 4.9 percentage points from 1970 to 2010.
- The rate at which the central city share of both population and total employment decreased slowed over time. 67% of the decrease in central city population share and 80% of the decrease in central city employment share took place in the 1970s and 1980s.
- The ratio of central city median household income to metropolitan median household income fell by 10.2 percentage points between 1970 and 2010.
- As was the case for population and employment, the rate at which the income ratio decreased slowed over time. 70% of the decrease in the central city-metropolitan area income ratio took place in the 1970s and 1980s.
- There was a significant increase in the number of cities with income growth that exceeded their metropolitan area's income growth. In the 1970s, there were 21 cities with income growth that exceeded the income growth in their metropolitan areas. This number fell to 17 in the 1980s before increasing to 27 in the 1990s and 34 in the 2000s.
- The gap in educational attainment between central cities and their metropolitan areas, measured by the percentage of residents aged 25 and over with at least a bachelor's degree, also increased over time. In the 1970s, the gap between central city educational attainment and metropolitan education attainment was 0.5 percentage points. By 2010, it had increased to 2.1 percentage points. The average metropolitan area experienced an increase of 17.3 percentage points in its educational attainment while the average central city experienced an increase of 15.7 percentage points.
- Occupation is the one area where central city growth exceeded metropolitan growth. While the percentage of workers in professional and executive occupations was lower in central cities for every year, the change in the percentage of workers in these occupations was slightly greater for central cities (10.9 percentage points) than for metropolitan areas (10.7 percentage points).

Notes

1 The SOCDS does not include observations for 2010. The latest data included in the SOCDS is from the 2008 American Community Survey. These values are used as a proxy for 2010 in the tables included in this chapter.
2 There were 19 cities with missing occupational data in the State of the Cities Database for 2010.

Bibliography

Mieszkowski, Peter and Edwin S. Mills (1993). "The Causes of Metropolitan Suburbanization", *Journal of Economic Perspectives*, 7(3): 135–147.
State of the Cities Data System, U.S. Department of Housing and Urban Development (1970, 1980, 1990, 2000, 2006). https://www.huduser.gov/portal/datasets/socds.html.

4

INCOME GENTRIFICATION IN U.S. CITIES, 1970–2010

Introduction

This chapter is the first of three consecutive chapters that attempt to measure the frequency of gentrification in U.S. cities from 1970 to 2010. Each chapter focuses on a single dimension of gentrification (income in this chapter, education in Chapter 5, and occupation in Chapter 6). Finally, Chapter 7 analyzes the various combinations of these three types of gentrification and identifies which combinations were the most common and how the most prevalent forms of gentrification change over time.

This study uses the Neighborhood Change Database (NCDB) to identify gentrifying neighborhoods in the United States from 1970 to 2010. The NCDB is ideal for this purpose because it provides U.S. Census data for each census year for census tracts with geographically consistent boundaries. This is important because census tract boundaries change over time, primarily due to population growth within the tract that leads to the tract being split into multiple tracts. The NCDB ensures that the same geographic area is being analyzed from year to year.

The initial sample for this study includes the core-based statistical areas (CBSAs) of the 100 largest cities by population in 1970. To analyze whether gentrification trends differ between large cities and small- and medium-sized cities, the full sample is split into a large city subsample and a small/medium city subsample. The large city subsample contains all cities with at least 100 census tracts in the central city. The large city subsample contains 47 cities in the 1970s and 1980s, 48 cities in the 1990s, and 49 cities in the 2000s.

To increase the likelihood that only residential tracts are being studied, tracts with a total population of less than 50 people and those with zero total housing units in the beginning year of each decade are deleted from the sample.[1] Because

DOI: 10.1201/9781003217459-4

TABLE 4.1 Number of Central City Tracts by Decade

		1970	1980	1990	2000	Regional Distribution			
						1970	1980	1990	2000
Full Sample	Total	15,122	15,329	15,381	15,337				
	Northeast	3,457	3,474	3,479	3,472	22.86%	22.66%	22.62%	22.64%
	Midwest	3,525	3,529	3,532	3,538	23.31%	23.02%	22.96%	23.07%
	South	4,677	4,795	4,808	4,819	30.93%	31.28%	31.26%	31.42%
	West	3,463	3,531	3,562	3,508	22.90%	23.03%	23.16%	22.87%
Large	Total	11,887	12,071	12,213	12,208				
	Northeast	2,809	2,823	2,827	2,824	23.63%	23.39%	23.15%	23.13%
	Midwest	2,614	2,618	2,620	2,625	21.99%	21.69%	21.45%	21.50%
	South	3,511	3,620	3,629	3,736	29.54%	29.99%	29.71%	30.60%
	West	2,953	3,010	3,137	3,023	24.84%	24.94%	25.69%	24.76%
Small/ Medium	Total	3,235	3,258	3,168	3,129				
	Northeast	648	651	652	648	20.03%	19.98%	20.58%	20.71%
	Midwest	911	911	912	913	28.16%	27.96%	28.79%	29.18%
	South	1,166	1,175	1,179	1,083	36.04%	36.07%	37.22%	34.61%
	West	510	521	425	485	15.77%	15.99%	13.42%	15.50%

of this, the total number of census tracts included in the sample differs from decade to decade. An alternate approach would be to only include those tracts that meet the criteria for every decade. This approach is rejected to allow for including the largest number of tracts in each decade.

Table 4.1 contains basic information regarding the size of the sample and the regional distribution of tracts for the full sample and the two subsamples. The full sample contains slightly more than 15,000 central city tracts in each decade. Slightly less than 80% of the central city tracts are in large cities. The full sample is somewhat concentrated in the South with 31% of the central city tracts in Southern cities. The remaining tracts are very evenly distributed across the other three census regions. The large city subsample has a higher proportion of tracts in the Northeast and West than the full sample while the small/medium city subsample has a higher proportion of tracts in the Midwest and South.

Results for Gentrifiable Tracts

Identifying gentrifying tracts is a two-step process. At its most basic level, gentrification is the process by which low-income, central city neighborhoods move up the metropolitan neighborhood hierarchy of socioeconomic status (SES), primarily through the in-migration of higher SES residents. This means that before a neighborhood can gentrify, it must be *gentrifiable*. For the purposes of this study, gentrifiable tracts are defined as central city census tracts with incomes that are less than 80% of the median income of all tracts in the CBSA containing the central city.

In addition, *low-income gentrifiable tracts* (defined as those with incomes between 50 and 80% of the area median income) and *very low-income gentrifiable tracts* (those with incomes less than or equal to 50% of the area median income) will be analyzed separately. Breaking down the results in this way will make it possible to determine the extent to which gentrification activity is concentrated in the gentrifiable neighborhoods with the highest incomes. It will also make it possible to determine whether gentrification trends differ between the two types of neighborhoods.

Neighborhood income is identified as the average household income (AHI) for each census tract. While median household income is a preferred measure, it is not reported in the NCDB for 1970 and 1980 while AHI is reported for every year. Because of this, AHI is used to measure neighborhood income in this study. Once again, the metropolitan median income for each year is calculated as the median AHI for all tracts in the CBSA containing the city.

Table 4.2 contains the results regarding the number of central city neighborhoods that are classified as gentrifiable for the full sample at the beginning of each decade included in the study. The regional distribution of gentrifiable tracts is provided as well as the split between low-income and very low-income gentrifiable tracts.[2]

In 1970, there were 5,280 gentrifiable census tracts in the full sample. This means that approximately 35% of central city tracts had the potential to gentrify. The percentage of central city tracts that are gentrifiable is slightly higher in the Northeast (39%) and Midwest (41%) and lower in the South (29%) and West (32%).

During the 1970s, the number of gentrifiable tracts increased by 25% to 6,593 and the percentage of central city tracts considered to be gentrifiable increased to

TABLE 4.2 Number of Gentrifiable Tracts by Decade

		1970	1980	1990	2000	Regional Distribution			
						1970	1980	1990	2000
Full	Total	5,280	6,593	6,963	7,185				
	Northeast	1,350	1,776	1,791	1,916	25.57%	26.94%	25.72%	26.67%
	Midwest	1,466	1,738	1,933	1,874	27.77%	26.36%	27.76%	26.08%
	South	1,355	1,758	1,875	1,956	25.66%	26.66%	26.93%	27.22%
	West	1,109	1,321	1,364	1,439	21.00%	20.04%	19.59%	20.03%
Large	Total	4,169	5,199	5,480	5,688				
	Northeast	1,026	1,376	1,383	1,476	24.61%	26.47%	25.24%	25.95%
	Midwest	1,155	1,359	1,470	1,428	27.70%	26.14%	26.82%	25.11%
	South	1,018	1,326	1,406	1,535	24.42%	25.50%	25.66%	26.99%
	West	970	1,138	1,221	1,249	23.27%	21.89%	22.28%	21.96%
Small/ Medium	Total	1,111	1,394	1,483	1,497				
	Northeast	324	400	408	440	29.16%	28.69%	27.51%	29.39%
	Midwest	311	379	463	446	27.99%	27.19%	31.22%	29.79%
	South	337	432	469	421	30.33%	30.99%	31.63%	28.12%
	West	139	183	143	190	12.51%	13.13%	9.64%	12.69%

43% by 1980. While the number of gentrifiable tracts increased in every census region, the increase was largest in the Northeast (31%) and the South (30%). Since neighborhoods are classified as gentrifiable based on how their incomes compared to the area median income, the large increase in the percentage of central city tracts considered to be gentrifiable reflects the extent to which central city income growth lagged suburban income growth in U.S. cities during the 1970s.

The number of gentrifiable tracts continued to increase during the 1980s and 1990s but the increases were much smaller than those that occurred during the 1970s. During the 1980s, the number of gentrifiable tracts increased by less than 6% to 6,963 and the percentage of central city tracts considered to be gentrifiable increased to 45%. During the 1990s, the number of gentrifiable tracts increased by slightly more than 3% to 7,185, representing almost 47% of the central city tracts in the sample. Overall, there was a 36% increase in the number of gentrifiable tracts between 1970 and 2000. The biggest increases were in the South (44%) and the Northeast (41%). The number of gentrifiable tracts in the West increased by 30%, while the Midwest had the smallest increase at 28%.

Table 4.2 also provides the breakdown of gentrifiable tracts between large cities and small/medium cities. As was the case for the distribution of tracts between the two subsamples, slightly less than 80% of the gentrifiable tracts are in the large city subsample in 1970 and the percentage remains stable over time. This suggests that there are no significant differences between large cities and small/medium cities with respect to the percentage of central city tracts considered to be gentrifiable. This conclusion carries over to the regional results as well in that there are no large differences between the percentage of central city tracts in the large city subsample for each region and the percentage of gentrifiable tracts in large cities.

Table 4.3 shows the five cities with the highest percentage of central city tracts considered to be gentrifiable at the beginning of each decade. In 1970, 1980, and 2000, two cities in Connecticut (Hartford and Bridgeport) had the highest percentage of their central city tracts classified as gentrifiable. In 1990, Bridgeport had the

TABLE 4.3 Cities with the Highest Percentage of Central City Tracts Classified as Gentrifiable

1970		1980		1990		2000	
Hartford, CT	79.5%	Bridgeport, CT	89.5%	Bridgeport, CT	100.0%	Bridgeport, CT	100.0%
Bridgeport, CT	76.3%	Hartford, CT	79.5%	Gary, IN	96.8%	Hartford, CT	92.3%
Cleveland, OH	67.2%	Newark, NJ	79.1%	Newark, NJ	83.9%	Newark, NJ	92.0%
Newark, NJ	66.3%	Paterson, NJ	75.8%	Hartford, CT	82.5%	Gary, IN	90.3%
Minneapolis, MN	63.8%	Cleveland, OH	72.9%	Cleveland, OH	78.5%	St. Louis, MO	76.4%

highest percentage while Hartford was fourth. Bridgeport also had the distinction of having 100% of its central city tracts classified as gentrifiable in both 1990 and 2000. Every city on the list is drawn from either the Northeast or the Midwest. There are no cities from the South and West among the top five most gentrifiable cities in any decade. There are three cities (Hartford, Bridgeport, and Newark) that are among the most gentrifiable cities in every decade.

To summarize, there has been a consistent upward trend in the number of gentrifiable tracts in U.S. cities. This trend holds for all census regions and for both large and small/medium cities. However, for the full sample, 68% of the total increase occurred in the 1970s, 19% took place in the 1980s, and the remaining 12% occurred in the 1990s. Thus, a more precise description of the trend is that there was a large increase in the number of gentrifiable tracts in the 1970s followed by relatively small increases in the two subsequent decades.

Measuring Gentrification

Once the set of central city tracts that have the potential to gentrify has been identified, the next step is to identify the gentrifiable tracts that experience gentrification. This is not an easy or straightforward task. There is no well-established or accepted definition of how to identify gentrifying neighborhoods using quantitative data. In addition, a purely quantitative approach only captures a portion of the neighborhood changes generated by gentrification. A purely quantitative approach necessarily operates from a basic premise that, at its most basic level, gentrification represents the upward mobility of lower-income central city neighborhoods in the metropolitan hierarchy of neighborhoods with respect to the SES of their residents. Thus, gentrifying neighborhoods are identified as gentrifiable neighborhoods that exhibit this type of upward mobility.

This study will identify gentrifying neighborhoods using three neighborhood indicators commonly associated with gentrification: income growth, the percentage of residents with a college degree, and the percentage of residents employed in professional and executive occupations.[3] This chapter will analyze income gentrification in U.S. cities during the four decades from 1970 to 2010. The next two chapters will then analyze educational and occupational gentrification over the same period. A final chapter will pull together the results on income, educational, and occupational gentrification and will identify how common the various combinations of these three types of gentrification are and whether the form that gentrification takes varies across time and across cities.

Gentrifying neighborhoods are identified using the same methodology for each of the three indicators. First, as mentioned above, only gentrifiable neighborhoods can be classified as gentrifying. Second, gentrifiable neighborhoods are classified as *slowly gentrifying* if their change in a decade for the indicator of interest is greater than the CBSA-level change in the same indicator. Third, gentrifiable neighborhoods are classified as *rapidly gentrifying* for a particular indicator if their decadal change in the indicator of interest exceeds the CBSA-level change in the indicator by at least 50%. The rationale for using two different measures of gentrification is to identify gentrifying neighborhoods using a relatively loose measure (slowly gentrifying) and a much stricter measure (rapidly gentrifying). Thus, the two measures can be thought of as providing an upper- and a lower-bound on the number of gentrifying neighborhoods in each decade.

As was mentioned in the introduction, a major contribution of this study is to apply a consistent measure of gentrification to a large sample of U.S. cities across several decades. Using a consistent measure over multiple decades makes it possible to clearly identify trends regarding the amount of gentrification over time. An additional advantage of the methodology is that applying the same criteria to multiple indicators of gentrification makes it possible to determine whether the *nature* of gentrification has changed over time. One decade may be characterized by higher levels of educational gentrification, while the next decade may see more income gentrification. Thus, in addition to the primary goal of documenting the *amount* of gentrification over time, this study seeks, by using multiple indicators of gentrification, to determine whether there are changes in the specific forms that gentrification takes from city to city and from decade to decade.

Income Gentrification

One of the more common ways of thinking about gentrification is that it is a process in which middle- and upper-income households move into previously lower-income neighborhoods. Because of this, it makes sense to begin by analyzing income growth in gentrifiable neighborhoods. The rest of this chapter will document the levels and trends in income gentrification in U.S. cities from 1970 to 2010.

As was mentioned before, neighborhood income is measured using the AHI for each year. Metropolitan income is measured as the median AHI for all census tracts in the CBSA. Income growth in each decade is measured by the absolute change in neighborhood or metropolitan income during the decade.[4] A gentrifiable neighborhood is classified as slowly gentrifying with respect to income if the neighborhood's change in income during a decade is simply greater than the metropolitan change in income. A gentrifiable neighborhood is classified as rapidly gentrifying with respect to income if its income change during a decade is at least 50% greater than the metropolitan change in income.

Table 4.4 contains the results regarding the number of tracts that are identified as income gentrifying during each decade. The results are provided for all gentrifiable tracts as well as separately for low-income and very low-income gentrifiable tracts. Focusing first on the results for all gentrifiable tracts reveals that the amount of income gentrification increased every decade from the 1970s to the 2000s. The largest increase was from the 1970s to the 1980s where the number of income-gentrifying tracts almost tripled. There was a 90% increase from the 1980s to the 1990s and a 39% increase from the 1990s to the 2000s. So, while the number of

TABLE 4.4 Number of Income-Gentrifying Tracts by Decade

	Number of Tracts that Income-Gentrify			
Total Gentrifying	1970s	1980s	1990s	2000s
All Gentrifiable Tracts	222	644	1,221	1,691
Low-Income Gentrifiable	208	356	944	1,412
Very Low-Income Gentrifiable	14	288	277	279
Slowly Gentrifying				
All Gentrifiable Tracts	178	357	963	875
Low-Income Gentrifiable	171	305	744	751
Very Low-Income Gentrifiable	7	52	219	124
Rapidly Gentrifying				
All Gentrifiable Tracts	44	287	258	816
Low-Income Gentrifiable	37	51	200	661
Very Low-Income Gentrifiable	7	236	58	155

income-gentrifying tracts reached its peak in the 2000s, the largest increase in income gentrification relative to the previous decade occurred in the 1980s.

Since a prerequisite for a neighborhood to gentrify is that the neighborhood must first be gentrifiable, another way of measuring how much gentrification is occurring is to calculate the percentage of gentrifiable neighborhoods that gentrify. These results are provided in Table 4.5. For all gentrifiable tracts, the percentage of gentrifiable tracts that income gentrified essentially doubled from the 1970s to the 1980s and, once again, from the 1980s to the 1990s. The increase slowed from the 1990s to the 2000s so that during the 2000s, 24% of all gentrifiable tracts gentrified by income in some way. This, of course, supports the previous result that income gentrification increased each decade and reached its highest level in the 2000s.

TABLE 4.5 Percentage of Gentrifiable Tracts that Income Gentrify by Decade

	Percentage of Gentrifiable Tracts That Income Gentrify			
Total Gentrifying	1970s	1980s	1990s	2000s
All Gentrifiable Tracts	4.20%	9.77%	17.54%	23.58%
Low-Income Gentrifiable	4.50%	6.98%	18.07%	24.16%
Very Low-Income Gentrifiable	2.11%	19.28%	15.92%	21.02%
Slowly Gentrifying				
All Gentrifiable Tracts	3.37%	5.41%	13.83%	12.20%
Low-Income Gentrifiable	3.70%	5.98%	14.24%	12.85%
Very Low-Income Gentrifiable	1.06%	3.48%	12.59%	9.34%
Rapidly Gentrifying				
All Gentrifiable Tracts	0.83%	4.35%	3.71%	11.38%
Low-Income Gentrifiable	0.80%	1.00%	3.83%	11.31%
Very Low-Income Gentrifiable	1.06%	15.80%	3.33%	11.68%

Looking now at the overall results for low- and very low-income gentrifiable tracts reveals that the surge in income gentrification during the 1980s was concentrated in very low-income gentrifiable tracts. There was an almost 20-fold increase in the number of income-gentrifying tracts in very low-income gentrifiable tracts in the 1980s relative to the 1970s. For sake of comparison, the number of income-gentrifying tracts in low-income tracts increased by 71% relative to the 1970s. However, the results for the subsequent decades show that the surge in income gentrification in very low-income gentrifiable tracts was short-lived. The number of income-gentrifying tracts in this income band fell by 4% in the 1990s relative to the 1980s and then increased only 1% in the 2000s relative to the 1990s. The number of income-gentrifying tracts in low-income gentrifiable tracts during the 1990s was more than 2.5 times greater than the number in the 1980s. In addition, there was a 50% increase in the 2000s relative to the 1990s. Thus, the results show that, except for the 1980s, income gentrification was heavily concentrated in low-income gentrifiable tracts. In the 1970s, 94% of income-gentrifying tracts were low-income gentrifiable tracts. The percentage fell to 55% during the 1980s and then increased to 77% in the 1990s and 84% in the 2000s.

The results regarding the percentage of low- and very low-income gentrifiable tracts that income gentrified support many of the conclusions drawn from analyzing the overall number of income-gentrifying tracts. The large increase in income gentrification in the 1980s relative to the 1970s was concentrated in very low-income gentrifiable tracts. While the probability that a low-income gentrifiable tract experienced income gentrification increased from 4.5% in the 1970s to 7.0% in the 1980s, the probability that a very low-income gentrifiable tract experienced income gentrification increased from 2.1% in the 1970s to 19.3% in the 1980s. During the 1980s, almost one-in-five very low-income gentrifiable tracts experienced income gentrification. However, as was noted when analyzing the number of gentrifying tracts, the surge during the 1980s quickly abated. During the 1990s, when the probability that a low-income tract experienced income gentrification jumped from 7.0% to 18.1%, the probability that a very low-income tract income gentrified fell from 19.3% to 15.9%. During the 2000s, the percentage of gentrifiable tracts that gentrified increased by approximately the same amount in both types of gentrifiable neighborhoods so that 24.2% of low-income and 20.8% of very low-income gentrifiable tracts experienced income gentrification in the 2000s.

The results from segmenting income gentrification into slow and rapid gentrification reveal some interesting patterns. In the 1970s, 80% of the income-gentrifying tracts gentrified slowly. This percentage fell to 55% in the 1980s as there was a surge in the number of tracts that rapidly gentrified. The percentage jumped to 79% in the 1990s and fell to 52% in the 2000s. In the 1970s and 1990s, income gentrification was primarily slow and in both the 1980s and 2000s, the number of tracts that rapidly gentrified increased substantially. For example, in the 1980s, the number of slowly gentrifying tracts essentially doubled relative to the 1970s, while the number of rapidly gentrifying tracts in the 1980s was more, six times greater, than the number in the 1970s. Similarly, the number of slowly gentrifying tracts in the 2000s fell by 9% relative to the 1990s, while the number of rapidly gentrifying tracts more than tripled in the 2000s when compared to the 1990s.

As was mentioned above, the increase in income gentrification during the 1980s was heavily concentrated in very low-income gentrifiable tracts. This was especially true for rapidly gentrifying tracts. The number of very low-income gentrifiable tracts that slowly gentrified was more than seven times higher in the 1980s than 1970s, while the number of rapidly gentrifying very low-income gentrifiable tracts was 33 times higher in the 1980s than 1970s. In the 1990s, the increase in slowly gentrifying tracts was most visible in the very low-income gentrifiable tracts. The number of slowly gentrifying very low-income tracts was more than four times higher in the 1990s relative to the 1980s, while the number of low-income tracts that slowly gentrified was almost 2.5 times higher in the 1990s than in the 1980s. Similarly, the decrease in rapidly gentrifying tracts in the 1990s was entirely due to the large decrease in the number of rapidly gentrifying very low-income tracts. There was a 75% decrease in the number of rapidly gentrifying very low-income tracts in the 1990s, while the number of rapidly gentrifying low-income tracts was almost four times greater in the 1990s than in the 1980s. Likewise, the decrease in slowly gentrifying tracts in the 2000s was also entirely due to a large decrease in the number of very low-income gentrifiable tracts that slowly gentrified. There was a 44% decrease in the number of slowly gentrifying very low-income tracts in the 2000s, while the number of slowly gentrifying low-income tracts increased very slightly. Finally, the surge in rapidly gentrifying tracts in the 2000s was spread almost equally across low- and very low-income tracts with the number of rapidly gentrifying low-income tracts more than tripling relative to the 1990s, while the number of rapidly gentrifying very low-income tracts was 2.7 times higher in the 2000s than the number in the 1990s.

The results regarding the percentage of gentrifiable tracts that experience either slow or rapid income gentrification illuminate several of the previous results as well. During the 1970s, low-income gentrifiable tracts were almost five times as likely to slowly gentrify as very low-income tracts and both types of gentrifiable tracts were equally likely to rapidly gentrify. As was noted above, during the 1980s, there was a very large increase in the likelihood that a very low-income gentrifiable neighborhood income gentrified and that the increase was primarily due to a large increase in the probability that such neighborhoods rapidly gentrified. The probability that a very low-income gentrifiable tract rapidly gentrified increased from 1.1% in the 1970s to 15.8% in the 1980s. For sake of comparison, the probability that a very low-income gentrifiable tract slowly gentrified increased from 1.1% to 3.5%. During the 1990s, the probability that a gentrifiable tract slowly gentrified increased substantially. For all gentrifiable tracts, that probability increased from 5.4% in the 1980s to 13.8% during the 1990s. For low-income gentrifiable tracts, the probability increased from 6.0% to 14.2%, while for very low-income gentrifiable tracts, it increased from 3.5% to 12.6%. The percentage of all gentrifiable tracts that rapidly gentrified fell from 4.4% during the 1980s to 3.7% during the 1990s. The decrease was due to a large decrease in the percentage of very low-income gentrifiable tracts that rapidly gentrified. For very low-income gentrifiable tracts, the percentage that rapidly gentrified fell from 15.8% in the 1980s to 3.3% in the 1990s. The percentage of low-income gentrifiable tracts that rapidly gentrified increased from 1.0% in the 1980s to 3.8% in the 1990s. During the 2000s, the percentage of all gentrifiable tracts that slowly gentrified decreased from 13.8% in the 1990s to 12.2% in the 2000s. The decrease was larger for very low-income gentrifiable tracts where the percentage dropped from 12.6% in the 1990s to 9.2% in the 2000s. For low-income gentrifiable tracts, the percentage that slowly gentrified decreased from 14.2% in the 1990s to 12.9% in the 2000s. The percentage of all gentrifiable tracts that rapidly gentrified increased from 3.7% in the 1990s to 11.4% in the 2000s. This large increase mirrors the results reported previously regarding the very large increase in the number of tracts that experienced rapid income gentrification in the 2000s relative to the 1990s. Both low-income and very low-income gentrifiable tracts experienced comparable increases in the percentage of tracts that rapidly gentrified. For low-income gentrifiable tracts, the percentage increased from 3.8% in the 1990s to 11.3% in the 2000s, while the percentage for very low-income gentrifiable tracts increased from 3.3% in the 1990s to 11.6% in the 2000s.

Table 4.6 breaks down the results by census region. Looking first at the overall income gentrification results reveals that the decade-by-decade increase in the number of income-gentrifying tracts was experienced by every region in every decade. During the 1980s, the number of income-gentrifying tracts relative to the 1970s ranged from an increase of 170% in the South to an increase of 261% in the Northeast. During the 1990s, the increase ranged from a 49% increase in the

TABLE 4.6 Number of Income-Gentrifying Tracts by Census Region

	Number of Income-Gentrifying Tracts			
Total Gentrifying	1970s	1980s	1990s	2000s
Northeast	36	130	256	424
Midwest	58	163	356	427
South	84	227	339	460
West	44	124	270	380
Slowly Gentrifying				
Northeast	31	109	200	250
Midwest	53	69	287	179
South	57	111	263	205
West	37	68	213	241
Rapidly Gentrifying				
Northeast	5	21	56	174
Midwest	5	94	69	248
South	27	116	76	255
West	7	56	57	139

South to a 118% increase in the Midwest. Finally, the increase in the 2000s relative to the 1990s ranged from a 20% increase in the Midwest to a 66% increase in the Northeast.

Decomposing the overall gentrification levels into slow and rapid gentrification reveals some interesting regional differences in income gentrification trends. During the 1970s, except for the South, almost all income gentrification was slow gentrification. In the South, 68% of the income-gentrifying tracts were slowly gentrifying, the percentage in the other three regions ranged from 84% in the West to 91% in the Midwest. The surge in rapid income gentrification during the 1980s that was documented above was experienced in only three of the four regions. In the Northeast, only 16% of the income gentrification that occurred in the 1980s was rapid gentrification. In the Midwest, 58% of the gentrification was rapid, while 51% of the income gentrification in the South and 45% of the income gentrification in the West were rapid. In the 1990s, there was a return in all four regions to a majority of the income gentrification being slow gentrification. In the 1990s, the range was from 78% in the South to 81% in the Midwest. During the 1990s, the split between slow and rapid gentrification was almost identical across all four regions. Finally, the second surge in rapid gentrification in the 2000s that was documented above was experienced by all four regions. The percentage of income-gentrifying tracts that rapidly gentrified ranged from 37% in the West to 59% in the Midwest. Thus, unlike the 1980s where the Northeast did not experience a surge in rapid income gentrification, the 2000s saw a large increase in the amount of rapid income gentrification in all four census regions.

Table 4.7 provides the regional breakdowns in the percentage of gentrifiable tracts that income gentrified during each decade and identifies some interesting

TABLE 4.7 Percentage of Gentrifiable Tracts that Income Gentrify by Census Region

	Percentage of Gentrifiable Tracts that Income-Gentrify			
Total Gentrifying	1970s	1980s	1990s	2000s
Northeast	2.67%	7.32%	14.29%	22.13%
Midwest	3.96%	9.38%	18.42%	22.79%
South	6.20%	12.91%	18.08%	23.52%
West	3.97%	9.39%	19.79%	26.67%
Slowly Gentrifying				
Northeast	2.30%	6.14%	11.17%	13.05%
Midwest	3.62%	3.97%	14.85%	9.55%
South	4.21%	6.31%	14.03%	10.48%
West	3.34%	5.15%	15.62%	16.91%
Rapidly Gentrifying				
Northeast	0.37%	1.18%	3.13%	9.08%
Midwest	0.34%	5.41%	3.57%	13.23%
South	1.99%	6.60%	4.05%	13.04%
West	0.63%	4.24%	4.18%	9.75%

regional trends. The total gentrification results for the 1970s show that a much higher percentage of gentrifiable tracts in Southern cities income gentrified than in any of the other regions. While 6.2% of all gentrifiable tracts in Southern cities experienced some form of income gentrification, approximately 4.0% of Midwestern and Western cities and 2.7% of Northeastern gentrifiable tracts experienced income gentrification in the 1970s. The percentage increased in all four regions in the 1980s relative to the 1970s. The largest increase was in the South where the percentage increased from 6.2% in the 1970s to 12.9% in the 1980s. As was the case in the 1970s, gentrifiable tracts in Southern cities were more likely to income gentrify than those in any of the other regions. Both Midwestern and Western cities saw the percentage of gentrifiable tracts that income gentrified increase from 4.0% in the 1970s to 9.4% in the 1980s. The smallest increase in the 1980s relative to the 1970s was for Northeastern cities where the percentage increased from 2.7% in the 1970s to 7.3% in the 1980s. Gentrifiable tracts in Northeastern cities remained the least likely to income gentrify during the 1980s. Once again, the percentage of gentrifiable tracts that experienced income gentrification increased for all four regions in the 1990s relative to the 1980s. This time the largest increase was in the West where the overall percentage of gentrifiable tracts that experienced income gentrification increased from 9.4% in the 1980s to 19.8% in the 1990s. The next largest increase was for Midwestern cities where the percentage increased from 9.4% in the 1980s to 18.4% in the 1990s. Northeastern cities saw their percentage almost double from 7.3% in the 1980s to 14.3% in the 1990s. Southern cities had the smallest increase in the 1990s relative to the 1980s. Their percentage increased from 12.9% in the 1980s to 18.1% in the 1990s. In the 1990s, the highest percentage of gentrifiable tracts that experienced gentrification was for Western cities at 19.8%, while the lowest percentage was still for Northeastern cities at 14.3%. During the 2000s, the percentage of gentrifiable tracts that income gentrified once again increased for all four census regions. The largest increase was for Northeastern cities where the percentage increased from 14.3% in the 1990s to 22.1% in the 2000s. The smallest increase was for Midwestern cities where the percentage increased from 18.4% to 22.8%. During the 2000s, the highest percentage of gentrifiable tracts that gentrified was for Western cities at 26.4% and the smallest percentage was, once again, Northeastern cities at 22.1%. Northeastern cities had the lowest percentage of gentrifiable tracts that income gentrified in all four decades.

The results that separately identify the percentage of gentrifiable tracts that slowly and rapidly income-gentrified also highlight some interesting regional differences. In the 1970s, the South had the highest percentage of gentrifiable tracts that both slowly and rapidly gentrified, while the Northeast had the smallest percentage of slowly gentrifying tracts and the Midwest had the smallest percentage of rapidly gentrifying tracts. The percentages for the 1980s were higher for all four regions for both types of gentrification. The largest increase in the percentage of gentrifiable tracts that slowly gentrified was in the Northeast where the percentage increased from 2.3% to 6.1%. The smallest increase was in the Midwest where the percentage increased from 3.6% to 4.0%. The highest percentage of gentrifiable tracts that slowly gentrified was in the South at 6.3%, while the smallest percentage was 4.0% in the Midwest. For rapidly gentrifying tracts, the largest increase in the 1980s relative to the 1970s was for the Midwest where the percentage increased from 0.3% to 5.4% and the smallest increase was in the Northeast where the percentage increased from 0.4% to 1.2%. The highest percentage was 6.6% in the South and the lowest percentage was 1.2% in the Northwest. During the 1990s, the percentage of gentrifiable tracts that slowly gentrified increased relative to the 1980s in all four census regions. The largest increase was in the Midwest where the percentage increased from 4.0% in the 1980s to 14.9% in the 1990s. The smallest increase was in the Northeast where the percentage increased from 6.1% to 11.2%. The highest percentage in the 1990s was in the West at 15.6% and the lowest percentage was 11.2% in the Northeast. The results for rapidly gentrifying tracts in the 1990s show that the percentage of gentrifiable tracts that rapidly gentrified decreased in three of the four regions. The only exception was the Northeast where the percentage increased from 1.2% in the 1980s to 3.1% in the 1990s. The largest decrease was in the South where the percentage fell from 6.6% to 4.1%. The percentages in the 1990s were very similar for all four regions with the lowest percentage being 3.1% in the Northeast and the highest being 4.2% in the West. Finally, in the 2000s, the percentage of gentrifiable tracts that slowly gentrified decreased in both the Midwest and the South. In the Midwest, the percentage fell from 14.9% to 9.6% and in the South, it fell from 14.0% to 10.5%. Among the other two regions, the percentage in the Northeast increased from 11.2% to 13.1% and the percentage in the West increased from 15.6% to 16.7%. The percentage of gentrifiable tracts that rapidly gentrified increased in all four regions in the 2000s relative to the 1990s. The biggest increase was in the Midwest where the percentage increased from 3.6% in the 1990s to 13.2%. The South also experienced a fairly large increase as its percentage jumped from 3.6% in the 1990s to 13.0% in the 2000s. The percentage for the Northeast increased from 3.1% to 9.1%, while the percentage for the West increased from 4.2% to 9.7%. These results mirror those regarding the number of rapidly gentrifying tracts in the 2000s which found a large increase in the number of rapidly gentrifying tracts in the 2000s relative to the 1990s. One consistent finding regarding income gentrification is that there was a large increase in the amount of rapid income gentrification during the 2000s relative to earlier decades.

Table 4.8 provides the results for large cities (those with more than 100 census tracts in the central city) and small/medium cities. Both large and small/medium cities saw total income gentrification levels increase in each decade. During the 1980s, the increase relative to the 1970s was greater for small/medium cities than for large cities. However, both types of cities saw the total number of income-gentrifying tracts more than double from the 1970s to the 1980s. During the 1990s, the increase relative to the 1980s was 95% for large cities and 70% for small/medium cities. Finally, the disparity between the two types of cities was greatest for the 2000s relative to the 1990s. The number of income-gentrifying tracts in large

TABLE 4.8 Number of Tracts that Income Gentrify by City Size

	Number of Tracts			
Total Gentrifying	1970s	1980s	1990s	2000s
Large Cities	184	502	979	1,417
Small/Medium Cities	38	142	242	282
Slowly Gentrifying				
Large Cities	148	287	763	738
Small/Medium Cities	30	70	200	141
Rapidly Gentrifying				
Large Cities	36	215	216	679
Small/Medium Cities	8	72	42	141

cities increased by 44% which was more than twice as large as the 17% increase for small/medium cities.

Decomposing the income-gentrifying tracts into slowly and rapidly gentrifying tracts shows that the large increase in rapid gentrification during the 1980s was somewhat more pronounced for small/medium cities than for large cities. The results for the 1990s show that the drop in the number of rapidly gentrifying tracts relative to the 1980s was entirely due to a 42% drop in small/medium cities. The number of rapidly gentrifying tracts in large cities was essentially constant across the two decades. Finally, the results for the 2000s reveal that the two types of cities had comparable experiences relative to the 1990s. Both types experienced a decrease in the number of slowly gentrifying tracts. Large cities experienced a 4% decrease while small/medium cities experienced a 30% decrease. In addition, both types of cities experienced an increase in rapidly gentrifying tracts that exceeded 200%. The surge in rapid income gentrification during the 2000s was evenly spread across large and small/medium cities.

The results in Table 4.9 reveal that, for the most part, city size does not appear to have a large effect on the percentage of gentrifiable tracts that income gentrify. During the 1970s, gentrifiable tracts in large cities were slightly more likely to

TABLE 4.9 Percentage of Gentrifiable Tracts that Income-Gentrify by City Size

	Percentage of Gentrifiable Tracts that Income Gentrify			
Total Gentrifying	1970s	1980s	1990s	2000s
Large Cities	4.41%	9.66%	17.86%	24.81%
Small/Medium Cities	3.42%	10.19%	16.32%	18.84%
Slowly Gentrifying				
Large Cities	3.55%	5.52%	13.92%	12.92%
Small/Medium Cities	2.70%	5.02%	13.49%	9.42%
Rapidly Gentrifying				
Large Cities	0.86%	4.14%	3.94%	11.89%
Small/Medium Cities	0.72%	5.16%	2.83%	9.42%

income gentrify than those in small/medium cities but the differences were small. Large city gentrifiable tracts had a 1 percentage point higher probability of income gentrifying, a 0.9 percentage point higher probability of slowly gentrifying, and a 0.1 percentage point large probability of rapidly gentrifying. During the 1980s, the percentage increased more rapidly in small/medium cities than large cities. The increase in small/medium cities was such that gentrifiable tracts in small/medium cities had a 0.5 percentage point higher probability of experiencing some income gentrification and a 1 percentage point higher probability of rapidly gentrifying. Large cities were still more likely to slowly gentrify. During the 1990s, both types of cities experienced comparable increases in the probability that they income gentrified in some form and the probability that they slowly gentrified. The only differences during the 1990s was that there was a larger decline in the percentage of tracts that rapidly gentrified in small/medium cities than in large cities. The 2000s were the only decade in which city size plays a role in the percentage of gentrifiable tracts that income gentrify. During the 2000s, large cities were 5.9 percentage points more likely to experience any income gentrification, 3.5 percentage points more likely to slowly gentrify, and 2.5 percentage points more likely to rapidly gentrify. Income gentrification in the 2000s appears to have been somewhat more concentrated in large cities when compared to previous decades.

City-level Income Gentrification

The final section of this chapter will turn to the question of which cities experienced the most income gentrification between 1970 and 2010. The amount of income gentrification will be estimated two ways. First, the percentage of central city tracts that are identified as income gentrifying is calculated. This percentage is calculated for overall gentrification as well as for slow and rapid gentrification. Second, since a tract cannot be identified as gentrifying unless it is first gentrifiable, the percentage of gentrifiable tracts that are identified as gentrifying is also calculated for each city. Both percentages are calculated as an overall percentage and separately for slow and rapid gentrification.

Table 4.10 contains the five cities with the highest amount of income gentrification in each decade for both measures of gentrification activity. Panel A provides the results using the percentage of central city tracts that income-gentrify as the measure of the extent of gentrification, while Panel B provides the results using the percentage of gentrifiable tracts that income-gentrify. During the 1970s, Des Moines and Albuquerque had the highest percentage of their central city tracts gentrify. Des Moines was the only city with more than 10% of its central city tracts gentrifying. Ignoring Virginia Beach, which had only two gentrifiable tracts in 1970, Des Moines and Albuquerque also had the highest percentages of their gentrifiable tracts gentrify. Looking at the results for slow vs. rapid gentrification reveals that all the gentrification in both Des Moines and Albuquerque was slow gentrification. As was indicated above, there was only a small amount of rapid

TABLE 4.10 Cities with the Highest Levels of Income Gentrification for Each Decade

Panel A: Percentage of Central City That Gentrifies

	1970s		1980s		1990s		2000s	
Overall	Des Moines, IA	10.5%	Jersey City, NJ	19.4%	Shreveport, LA	24.56%	Atlanta, GA	28.68%
	Albuquerque, NM	9.1%	Madison, WI	16.4%	Grand Rapids, MI	20.00%	St. Louis, MO	26.42%
	Santa Ana, CA	5.8%	Lincoln, NE	15.2%	Santa Ana, CA	19.23%	Flint, MI	24.39%
	Shreveport, LA	5.3%	Fort Worth, TX	13.9%	Denver, CO	17.61%	Washington, DC	23.60%
	Honolulu, HI	5.2%	Arlington, VA	13.6%	Miami, FL	17.17%	Denver, CO	23.40%
Slow	Des Moines, IA	10.5%	Arlington, VA	13.6%	Grand Rapids, MI	18.0%	St. Louis, MO	16.04%
	Albuquerque, NM	9.1%	Jersey City, NJ	13.4%	Santa Ana, CA	17.3%	Washington, DC	13.48%
	Santa Ana, CA	5.8%	Paterson, NJ	9.1%	Spokane, WA	14.5%	Portland, OR	12.41%
	New Orleans, LA	4.5%	Salt Lake City, UT	7.5%	Denver, CO	14.1%	Sacramento, CA	11.76%
	Chicago, IL	3.8%	Norfolk, VA	6.3%	Denver, CO	14.0%	Newark, NJ	11.49%
Rapid	Shreveport, LA	5.3%	Madison, WI	14.8%	Shreveport, LA	10.5%	Atlanta, GA	23.26%
	Houston, TX	2.7%	Lincoln, NE	13.6%	Miami, FL	5.1%	Denver, CO	16.31%
	Denver, CO	2.2%	Fort Worth, TX	11.9%	Flint, MI	4.9%	Cleveland, OH	12.99%
	Fort Lauderdale, FL	2.1%	Columbus, OH	9.1%	Warren, MI	4.9%	Austin, TX	10.99%
	Honolulu, HI	2.1%	Kansas City, MO	8.8%	Jersey City, NJ	4.5%	St. Louis, MO	10.38%

(Continued)

TABLE 4.10 (Continued)

Panel B: % of Gentrifiable Tracts that Income-Gentrify

	1970s		1980s		1990s		2000s	
	City	%	City	%	City	%	City	%
Total	Virginia Beach, VA	50.0%	Virginia Beach, VA	50.0%	Virginia Beach, VA	75.00%	Arlington, VA	50.00%
	Albuquerque, NM	36.7%	Greensboro, NC	36.7%	Anaheim, CA	75.00%	Santa Ana, CA	50.00%
	Des Moines, IA	28.6%	Lincoln, NE	28.6%	Santa Ana, CA	66.67%	Atlanta, GA	46.84%
	Honolulu, HI	20.8%	Anaheim, CA	20.8%	Shreveport, LA	60.87%	Seattle, WA	45.00%
	Santa Ana, CA	20.0%	Madison, WI	20.0%	Grand Rapids, MI	47.62%	Portland, OR	42.62%
Slow	Albuquerque, NM	36.7%	Arlington, VA	36.7%	Virginia Beach, VA	75.0%	Santa Ana, CA	37.50%
	Des Moines, IA	28.6%	Virginia Beach, VA	28.6%	Santa Ana, CA	60.0%	Seattle, WA	30.00%
	Santa Ana, CA	20.0%	Anaheim, CA	20.0%	Grand Rapids, MI	42.9%	Portland, OR	29.51%
	Honolulu, HI	12.5%	Jersey City NJ	12.5%	Spokane, WA	36.4%	Spokane, WA	27.78%
	Birmingham, AL	11.8%	Fort Lauderdale, FL	11.8%	Seattle, WA	35.3%	Corpus Christi, TX	27.27%
Rapid	Virginia Beach, VA	50.0%	Greensboro, NC	50.0%	Anaheim, CA	37.5%	Atlanta, GA	37.97%
	Shreveport, LA	16.7%	Lincoln, NE	16.7%	Shreveport, LA	37.5%	Arlington, VA	35.71%
	Fort Lauderdale, FL	14.3%	Madison, WI	14.3%	Tulsa, OK	33.3%	St. Petersburg, FL	33.33%
	Houston, TX	10.3%	Fort Worth, WI	10.3%	Riverside, CA	25.4%	Austin, TX	29.58%
	Honolulu, HI	8.3%	Riverside, CA	8.3%	El Paso, TX	25.0%	Denver, CO	29.11%

gentrification in the 1970s (a total of 44 tracts throughout the entire sample), the city with the most rapid income gentrification was Shreveport in which 5.3% of the central city tracts (3 out of 57) and 16.7% of the gentrifiable tracts (3 out of 18) experienced rapid income gentrification. In addition, 51 out of the 100 cities experienced some type of income gentrification during the 1970s with 46 having at least one tract slowly gentrify by income and 19 having at least one tract rapidly income gentrify.

The results for the 1980s show the impact of the higher levels of income gentrification in the 1980s relative to the 1970s. During the 1970s, there was only one city in which more than 10% of central city tracts experienced income gentrification. During the 1980s, nine cities had at least 10% of their central city tracts gentrify by income. Jersey City had the highest percentage of its central city tracts gentrify (19.4%), while Madison and Lincoln also had at least 15% of their central city tracts income gentrify. Arlington and Jersey City were the only two cities in which at least 10% of the central city slowly gentrified, while three cities (Madison, Lincoln, and Fort Worth) had at least 10% of their central city rapidly gentrify. There were four cities that had at least 40% of their gentrifiable tracts gentrify in some way: Virginia Beach (3 out of 6), Greensboro (7 out of 16), Lincoln (10 out of 24), and Anaheim (2 out of 5). Arlington and Virginia Beach were the only two cities that had at least 30% of their gentrifiable tracts slowly gentrify and Greensboro, Lincoln, and Madison were the only three cities in which at least 30% of their gentrifiable tracts rapidly gentrified. During the 1980s, the number of cities having at least one central city tract experience some sort of income gentrification increased to 85 compared to 51 in the 1970s. The number of cities with at least one slowly gentrifying tract increased from 46 to 77 and the number with at least one rapidly gentrifying tract increased from 19 to 62. Not only did the overall amount of gentrification increase in the 1980s relative to the 1970s but income gentrification also became more widespread.

The city-level results for the 1990s also point to an increase in the amount of income gentrification in U.S. cities relative to the 1980s. In the previous two decades, no city had experienced income gentrification in 20% or more of its central city tracts. During the 1990s, there were two cities in which 20% or more of the central city tracts experienced some sort of income gentrification. Shreveport had 24.5% (14 out of 57) of its central city tracts income gentrify, while Grand Rapids had exactly 20% (10 out of 50) experience income gentrification. In addition, there were 23 cities in which at least 10% of the central city tracts income gentrified. This is a significant increase from the 1980s when only nine cities had at least 10% of their central city tracts income gentrify. There were 14 cities in which at least 10% of the central city tracts slowly gentrified. Shreveport, at 10.5%, was the only city in which at least 10% of central city tracts rapidly gentrified. There were six cities in which at least 40% of the gentrifiable tracts experienced income gentrification during the 1990s. Among cities with at least ten gentrifiable tracts in 1990s, Santa Ana (67%) and Shreveport (61%) had the highest percentage of gentrifiable tracts experience income gentrification. There were three cities in which at least 40% of their gentrifiable tracts experienced slow income gentrification. Among cities with at least ten gentrifiable tracts, the highest percentage was for Santa Ana at 60%. Among cities with at least ten gentrifiable tracts, the highest percentage of gentrifiable tracts that experienced rapid income gentrification was for Shreveport at 26%. During the 1990s, 98 out of the 100 cities in the sample had at least one tract income gentrify in some way, 97 cities had at least one tract slowly gentrify, and 64 cities had at least one tract rapidly gentrify. Once again, in addition to the overall levels increasing in the 1990s relative to previous decades, there was also an increase in the number of cities experiencing income gentrification. This was especially true for slow income gentrification in which the number of cities having at least one slow gentrification tract increased from 77 in the 1980s to 97 in the 1990s.

Finally, the city-level results for the 2000s reveal higher levels of income gentrification when compared to earlier decades. In the 2000s, there were five cities in which at least 20% of central city tracts experience some sort of income gentrification. The highest percentage was in Atlanta where 28.7% of central city tracts experienced some sort of income gentrification. This is the highest percentage for any city between 1970 and 2010. St. Louis had 26.4% of its central city tracts income gentrify which also exceeded any of the values recorded for cities in the previous decade. Atlanta had 23.3% of its central city tracts rapidly gentrify which was also the highest value recorded in this study. In Denver, 16.3% of central city tracts rapidly gentrified which also exceeds any of the values from previous decades. St. Louis had the highest percentage of central city tracts that slowly gentrified at 16.0%. However, this was not the highest value recorded in the study. Grand Rapids and Santa Ana had higher percentages in the 1990s. During the 2000s, there were nine cities in which at least 40% of their gentrifiable tracts experienced some sort of income gentrification. This is the largest number in any of the decades. Among cities with at least ten gentrifiable tracts, Santa Ana (50%), Arlington (50%), and Atlanta (46.8%) had the highest percentages. However, while three cities (Virginia Beach, Santa Ana, and Grand Rapids) had at least 40% of their gentrifiable tracts slowly gentrify in the 1990s, there were no cities that had at least 40% of their gentrifiable tracts slowly gentrify in the 2000s. Santa Ana had the highest percentage at 37.5% (3 out of 8). There were three cities in which at least 30% of gentrifiable tracts rapidly gentrified. Atlanta had the highest percentage at 37.9%, while Arlington (35.7%) and St. Petersburg (33.3%) were also above 30%. Atlanta's percentage is the highest value for any of the decades included in the sample. During the 2000s, 94 out of the 99 cities had at least one gentrifiable tract income gentrify in some way. This is a slight decrease when compared to the 1990s but is still a high number. During the 2000s, 89 out of the 99 cities had at least one tract slowly gentrify and 88 out of the 99 cities had at least one tract rapidly gentrify. This means that there was a very large increase in the number of cities in which at least one tract rapidly gentrified (from 64 in the 1990s to 88 in the 2000s) and a decrease in the number of cities that slowly gentrified (from 97 in the 1990s to 89 in the 2000s). One of the key characteristics of income gentrification in the 2000s is that there was a substantial increase in the overall levels of rapid gentrification and in the number of cities that had tracts that were experiencing rapid income gentrification.

Conclusion

This chapter has analyzed income gentrification in U.S. cities from 1970 to 2010. Income-gentrifying census tracts were identified using a two-step process. First, a tract must be *gentrifiable* before it can gentrify. Gentrifiable tracts are defined as central city tracts with AHI less than 80% of the median AHI for all tracts in the CBSA. Second, a gentrifiable tract is classified as gentrifying if its income growth exceeds that of the CBSA. Two types of gentrifying tracts are identified. A gentrifiable tract is said to *slowly gentrify* if its change in AHI in a decade exceeds the change in the median AHI for the CBSA containing the tract. A gentrifiable tract is said to *rapidly gentrify* if its change in AHI exceeded the change in the median AHI for the CBSA by at least 50%.

The primary findings from the chapter can be summarized as:

- In 1970, 35% of central city census tracts were gentrifiable and the percentage grew in every subsequent decade so that, in 2000, almost 47% of central city tracts were gentrifiable.
- Most of the increase in the percentage of central city tracts that were gentrifiable occurred between 1970 and 1980. Almost 70% of the total increase occurred between 1970 and 1980.
- Income gentrification was rare in the 1970s with 1.5% of central city tracts and 4.2% of gentrifiable tracts gentrifying by income.
- The amount of income gentrification essentially doubled during the 1980s and 1990s, so that during the 1990s, 7.9% of central city tracts and 17.5% of gentrifiable tracts experienced income gentrification.
- The amount of income gentrification peaked in the 2000s but the increase in the 2000s relative to the 1990s was much smaller than the increases in the 1980s and 1990s.
- Rapid income gentrification was much more common in the 1980s and 2000s than in the 1970s and 1990s.
- In the 1980s, there was a very large increase in the number of very low-income-gentrifiable tracts that rapidly gentrified by income. In every other decade, rapid income gentrification was concentrated in low-income gentrifiable tracts.
- In the 1970s and 1980s, income gentrification was more common in cities in the South than cities in other Census regions.
- In the 1990s and 2000s, income gentrification was most common in cities in the West.
- Analysis of individual cities reveals that the number of cities affected by income gentrification increased as time went by and, also, the number of cities that experienced high levels of income gentrification also increased over time.

Notes

1 These inclusion criteria explain why additional cities are added to the large city sub-sample in the 1990s and 2000s. Sacramento is added to the large city subsample in the 1990s, while Tampa is added in the 2000s. These cities were added because they had additional tracts that met the criteria for inclusion in the sample and, therefore, the cities reached the threshold of having at least 100 central city tracts.
2 The results for the census regions are obtained by combining all the tracts in each region into a single sample and calculating the percentage of the central city tracts in each region that are classified as gentrifiable.
3 It would be useful to also analyze housing gentrification using median home values or median gross rents. However, these variables are not included in the NCDB for 1970 and 1980.
4 It should be emphasized that income change is measured by the actual change in income rather than by percentage change. The results were generated using both measures and actual changes were chosen as the preferred measure. Percentage changes tended to overidentify very low-income neighborhoods as gentrifying with respect to income even if they had relatively small changes in income. In the view of the author, following Mc-Kinnish et al. (2010), actual changes identify income-gentrifying neighborhoods more accurately.

Bibliography

McKinnish, Terra, Randall Walsh and Kirk T. White (2010). "Who Gentrifies Low-Income Neighborhoods?", *Journal of Urban Economics*, 67(2): 180–193.
Neighborhood Change Database (2013). Geolytics.

5

EDUCATIONAL GENTRIFICATION IN U.S. CITIES, 1970–2010

Introduction

Gentrification is most commonly associated with the movement of higher-income households into traditionally lower-income neighborhoods. Because of this, it was natural to analyze income gentrification in U.S. cities from 1970 to 2010 before moving on to other types of gentrification. However, there are dimensions to gentrification other than income and, in some cases, gentrification can involve the movement of highly educated but lower-income households into a gentrifiable neighborhood. This process, sometimes referred to as *studentification* (Rucks-Ahidiana 2020), suggests that studying the extent to which lower-income central city neighborhoods experienced rapid increases in the educational levels of their residents is also worthwhile. As was the case in the previous chapter, the ideal approach would involve studying the educational attainment of in-movers relative to the educational attainment of the original residents of the neighborhood. However, the data employed in this study does not allow for this which means that some of the neighborhoods identified as educationally gentrifying in this chapter may be experiencing increasing educational levels among current residents rather than the in-movement of more highly educated households. However, especially in the case of rapidly gentrifying neighborhoods, the magnitude of the change required for a neighborhood to be classified as gentrifying is more likely to be caused by the in-movement of more highly educated households rather than through a sudden large increase in the educational attainment of existing residents.

This chapter applies the same approach that was used in the previous chapter to analyze changes in neighborhood income to analyze changes in neighborhood education levels. Neighborhood educational attainment is measured by the percentage of neighborhood residents who are at least 25 years old and who have at

DOI: 10.1201/9781003217459-5

least a bachelor's degree. Gentrifiable neighborhoods that experience educational gentrification are identified by comparing the change in the percentage of neighborhood residents with at least a bachelor's degree to the change in the percentage of metropolitan residents with at least a bachelor's degree during each decade. Gentrifiable neighborhoods continue to be defined as central city census tracts with average household incomes (AHIs) that are less than 80% of the median AHI of all metropolitan tracts. A gentrifiable neighborhood is classified as experiencing slow educational gentrification if its change in educational attainment is simply greater than the change for the metropolitan area. A gentrifiable neighborhood in which the change in educational attainment is at least 50% greater than the metropolitan change is classified as experiencing rapid educational gentrification.

Table 5.1 contains the results regarding the total number of gentrifiable tracts that are identified as experiencing educational gentrification during each decade. The results are provided for all tracts that educationally gentrify as well as being broken down into the number that are slowly and rapidly gentrifying. The results are also provided for the number of gentrifying tracts that were low-income and very low-income gentrifiable tracts.

During the 1970s, 1,123 tracts are identified as experiencing some sort of educational gentrification. Of these, 58% experienced rapid gentrification and 90% of the tracts that gentrified were low-income gentrifiable tracts. The split between low-income and very low-income gentrifiable tracts is very similar for both slowly and rapidly gentrifying tracts: 90.4% of slowly gentrifying tracts and 90.6% of rapidly gentrifying tracts were in low-income gentrifiable tracts.

The number of tracts that experienced educational gentrification increased in the 1980s relative to the 1970s. Overall, the number of tracts experiencing some sort of educational gentrification increased by 49.6% to 1,680. The number of slowly gentrifying tracts increased by 31.9%, while the number of rapidly gentrifying

TABLE 5.1 Number of Educationally Gentrifying Tracts by Decade

	Number of Tracts that Educationally Gentrify			
Total Gentrifying	1970s	1980s	1990s	2000s
All Gentrifiable Tracts	1,123	1,680	1,782	2,914
Low-Income Gentrifiable	1,015	1,308	1,377	2,359
Very Low-Income Gentrifiable	108	372	405	555
Slowly Gentrifying				
All Gentrifiable Tracts	477	629	684	803
Low-Income Gentrifiable	430	512	514	649
Very Low-Income Gentrifiable	47	117	170	154
Rapidly Gentrifying				
All Gentrifiable Tracts	646	1,051	1,098	2,111
Low-Income Gentrifiable	585	796	863	1,710
Very Low-Income Gentrifiable	61	255	235	401

tracts increased by 62.7%. The percentage of gentrifying tracts that experienced rapid gentrification increased from 58% in the 1970s to 63% in the 1980s. There was a very large increase in the number of very low-income gentrifiable tracts that educationally gentrified. The number of educationally gentrifying tracts located in very low-income gentrifiable tracts increased by 244%. This includes a 149% increase in the number of slowly gentrifying tracts and a 318% increase in the number of rapidly gentrifying tracts. As was the case for income gentrification, there was a very large increase in the number of very low-income gentrifiable tracts that educationally gentrified in the 1980s when compared to the 1970s. In addition, the educational gentrification in very low-income tracts in the 1980s was predominantly rapid gentrification with almost 69% of the educational gentrification in these neighborhoods consisting of rapid gentrification. This is similar to what happened with income gentrification in the 1980s where 82% of the income gentrification in very low-income gentrifiable tracts in the 1980s was classified as rapid gentrification.

The number of educationally gentrifying tracts increased by only 6.1% in the 1990s relative to the 1980s. The number of tracts that slowly gentrified increased by 8.7% and the number that rapidly gentrified increased by 4.5%. The percentage of educationally gentrifying tracts that rapidly gentrified decreased from 63% in the 1980s to 62% in the 1990s. The number of educationally gentrifying tracts located in low-income gentrifiable tracts increased by 5.3%, while the number located in very low-income gentrifiable tracts increased by 8.9%. The two most interesting changes in the 1990s were that the number of slowly gentrifying tracts located in very low-income gentrifiable tracts increased by 45.3%, while the number of rapidly gentrifying tracts located in very low-income gentrifiable tracts decreased by 7.8%. This led to the percentage of educationally gentrifying very low-income tracts that rapidly gentrified decreasing from 69% in the 1980s to 58% in the 1990s.

Finally, there was another large increase in the amount of educational gentrification in the 2000s. The number of tracts experiencing some sort of educational gentrification increased by 63.5% in the 2000s relative to the 1990s. There was a 17.4% increase in the number of slowly gentrifying tracts and a 92.3% increase in the number of rapidly gentrifying tracts. The percentage of educationally gentrifying tracts that rapidly gentrified increased from 62% in the 1990s to 72% in the 2000s. There was a 71.3% increase in the total number of gentrifying tracts located in low-income gentrifiable tracts and a 37.0% increase in the number located in very low-income gentrifiable tracts. The largest increase in the 2000s was for rapidly gentrifying tracts located in low-income gentrifiable tracts. The number of tracts in this category increased by 98% relative to the 1990s. At the same time, there was a 9.4% decrease in the number of slowly gentrifying tracts located in very low-income tracts. So, while there was a large increase in the number of educationally gentrifying tracts in the 2000s relative to the 1990s, the increase was largest for rapidly gentrifying tracts located in low-income gentrifiable tracts.

TABLE 5.2 Percentage of Gentrifiable Tracts that Educationally Gentrify by Decade

	Percentage of Gentrifiable Tracts that Educationally Gentrify			
Total Gentrifying	1970s	1980s	1990s	2000s
All Gentrifiable Tracts	21.3%	25.5%	25.6%	40.6%
Low-Income Gentrifiable	22.0%	25.7%	26.4%	40.4%
Very Low-Income Gentrifiable	16.3%	24.9%	23.3%	41.8%
Slowly Gentrifying				
All Gentrifiable Tracts	9.0%	9.5%	9.8%	11.2%
Low-Income Gentrifiable	9.3%	10.0%	9.8%	11.1%
Very Low-Income Gentrifiable	7.1%	7.8%	9.8%	11.6%
Rapidly Gentrifying				
All Gentrifiable Tracts	12.2%	15.9%	15.8%	29.4%
Low-Income Gentrifiable	12.7%	15.6%	16.5%	29.3%
Very Low-Income Gentrifiable	9.2%	17.1%	13.5%	30.2%

Since, as has been mentioned before, the number of gentrifying tracts will be affected by the number of gentrifiable tracts, Table 5.2 uses the percentage of gentrifiable tracts that educationally gentrify to measure the amount of gentrification that takes place in each decade. During the 1970s, 21.3% of all gentrifiable tracts educationally gentrified. Low-income gentrifiable tracts were 5.7 percentage points more likely to gentrify than very low-income gentrifiable tracts. The percentage of gentrifiable tracts that gentrified increased in the 1980s relative to the 1970s for all categories. Overall, there was a 4.2-percentage point increase in the percentage of gentrifiable tracts that experienced some sort of educational gentrification. Very low-income gentrifiable tracts experienced an 8.6-percentage point increase in the percentage of the gentrified. The increase was large enough that in the 1980s, very low-income gentrifiable tracts were only 0.75 percentage points less likely to educationally gentrify than low-income gentrifiable tracts. Most of the increase in the probability that gentrifiable tracts gentrified was among rapidly gentrifying tracts. The probability that a gentrifiable tract rapidly gentrified increased by 3.7 percentage points, while the probability that it slowly gentrified only increased by 0.5 percentage points.

During the 1990s, there was a very slight increase in the percentage of gentrifiable tracts that experienced educational gentrification. The percentage increased by 0.11 percentage points during the 1990s relative to the 1980s. The percentage of low-income gentrifiable tracts that educationally gentrified increased by 0.71 percentage points, while the percentage of very low-income tracts that gentrified fell by 1.62 percentage points. The probability that a gentrifiable tract slowly gentrified increased by 0.28 percentage points and the probability that it rapidly gentrified fell by 0.17 percentage points.

Finally, there was a large increase in the percentage of gentrifiable tracts that educationally gentrified in the 2000s relative to the 1990s. Overall, the percentage of all gentrifiable tracts that experienced some sort of educational gentrification increased from 25.6% to 40.6%, a 15-percentage point increase. The percentage of low-income gentrifiable tracts that gentrified increased by 14.0 percentage points and the percentage of very low-income gentrifiable tracts that gentrified increased by 18.6 percentage points. Most of the increase in the percentage of gentrifiable tracts that gentrified was due to a very large increase in the percentage of gentrifiable tracts that rapidly gentrified. The percentage of gentrifiable tracts that rapidly gentrified increased by 13.7 percentage points in the 2000s relative to the 1990s, while the percentage of the slowly gentrified only increased by 1.4 percentage points. There was a very large increase in rapid educational gentrification during the 2000s such that almost 30% of all gentrifiable tracts rapidly gentrified during the 2000s.

Table 5.3 provides the regional breakdown of the tracts that educationally gentrified during each decade. The gentrifying tracts in the 1970s are very evenly distributed across regions. The Midwest had the most gentrifying tracts with 27.6% of the total, while the Northeast had the fewest with 20.0% of the total. The West had the largest number of slowly gentrifying tracts with 30.0% of the total, while the Northeast, with 19.5%, had the fewest. Finally, the Midwest had the most rapidly gentrifying tracts with 29.7% of the total, while the Northeast, with 20.3%, had the fewest. For the most part, educationally gentrifying tracts in the 1970s were evenly spread across the four census regions.

As was seen earlier, during the 1980s, there was a substantial increase in the amount of educational gentrification relative to the 1970s. Table 5.3 reveals that the biggest increase in educational gentrification was in the Northeast where the total number of gentrifying tracts increased by 117%, the number of slowly gentrifying

TABLE 5.3 Number of Educationally Gentrifying Tracts by Census Region

	Number of Educationally Gentrifying Tracts			
Total Gentrifying	1970s	1980s	1990s	2000s
Northeast	224	486	477	841
Midwest	310	410	416	671
South	302	391	444	740
West	287	393	445	662
Slowly Gentrifying				
Northeast	93	209	186	263
Midwest	118	169	195	198
South	123	130	159	187
West	143	121	144	155
Rapidly Gentrifying				
Northeast	131	277	291	578
Midwest	192	241	221	473
South	179	261	285	553
West	144	272	301	507

tracts increased by 125%, and the number of rapidly gentrifying tracts increased by 112%. There was also an 89% increase in the number of rapidly gentrifying tracts in the West. Interestingly, there was also a 15% decrease in the number of slowly gentrifying tracts in the West.

During the 1990s, there was only a 6% increase in the number of educationally gentrifying tracts. However, this aggregate change disguises some substantial differences between the regions. Both the South and the West saw the number of educationally gentrifying tracts increase by over 13%, while the number in the Midwest grew by only 1.5%. The number in the Northeast declined by 1.9%.

Finally, all four regions experienced significant growth in the number of educationally gentrifying tracts in the 2000s relative to the 1990s. The Northeast, the South, and the Midwest all had an increase of more than 60% in the number of gentrifying tracts. The largest increase was in the Northeast where the number increased by 76%. The West, with an increase of 48%, had the smallest increase. For all regions, the most significant increases were for rapid gentrification. The largest increase was in the Midwest where the number of rapidly gentrifying tracts increased by 114%. The Northeast experienced an increase of 99% in the number of educationally gentrifying tracts and the Midwest saw an increase of 94%. The smallest increase was for the West where the number of rapidly gentrifying tracts increased by 68%. The only significant increase in slow gentrification was for the Northeast where the number increased by 41%.

Table 5.4 contains the results concerning the percentage of gentrifiable tracts that gentrified educationally. During the 1970s, the West had the highest percentage of tracts that gentrified with 25.9% of its gentrifiable tracts experiencing

TABLE 5.4 Percentage of Gentrifiable Tracts that Educationally Gentrify by Census Region

	Percentage of Gentrifiable Tracts that Educationally Gentrify			
Total Gentrifying	1970s	1980s	1990s	2000s
Northeast	16.6%	27.4%	26.6%	43.9%
Midwest	21.2%	23.6%	21.5%	35.8%
South	22.3%	22.2%	23.7%	37.8%
West	25.9%	29.8%	32.6%	46.5%
Slowly Gentrifying				
Northeast	6.9%	11.8%	10.4%	13.7%
Midwest	8.1%	9.7%	10.1%	10.6%
South	9.1%	7.4%	8.5%	9.6%
West	12.9%	9.2%	10.6%	10.9%
Rapidly Gentrifying				
Northeast	9.7%	15.6%	16.3%	30.2%
Midwest	13.1%	13.9%	11.4%	25.2%
South	13.2%	14.9%	15.2%	28.3%
West	13.0%	20.6%	22.1%	35.6%

educational gentrification in some way. The lowest percentage was in the Northeast where only 16.6% of gentrifiable tracts experienced educational gentrification. During the 1980s, however, the Northeast had the largest increase in the percentage of gentrifiable tracts that gentrified relative to the 1970s with an increase of 10.8 percentage points. The West and the Midwest had much smaller increases at 3.9 and 2.4 percentage points, respectively, while the percentage in the South declined by 0.05 percentage points. The West still had the highest percentage of gentrifiable tracts that gentrified at 32.6%, while the Midwest had the lowest at 21.5%.

During the 1990s, the changes in the percentage of gentrifiable tracts that gentrified were much smaller than the changes in the 1980s. In fact, only two of the four regions saw their percentages increase. The percentages in the Midwest and Northeast decreased by 2.1 and 0.7 percentage points, while the largest increase was for the West at 2.9. The percentage for the South increased by 1.4 percentage points. The West still had the highest percentage of gentrifiable tracts that gentrified at 32.6%, while the South had the smallest percentage at 22.3%.

Finally, during the 2000s, all four regions experienced large increases in the percentage of gentrifiable tracts that educationally gentrified. The increases ranged from 13.8% in the West to 17.3% in the Northeast. The West once again had the highest percentage at 46.5%, while the lowest percentage was in the Midwest at 35.8%. The lowest percentage in the 2000s (35.8%) was higher than any of the percentages from the previous decades. The previous highest percentage was 32.6% for the West in the 1990s. As was the case for income gentrification, the highest levels of educational gentrification were in the 2000s. However, unlike income gentrification where the largest increases in gentrification activity were in the 1980s and 1990s, the largest increase in educational gentrification was also in the 2000s.

Table 5.5 contains the results regarding the amount of educational gentrification for large and small/medium cities. Recall that large cities are those with at least 100 census tracts in the central city. In the 1970s, 80.7% of all educationally gentrifying

TABLE 5.5 Number of Tracts that Educationally Gentrify by City Size

	Number of Tracts			
Total Gentrifying	1970s	1980s	1990s	2000s
Large Cities	906	1,307	1,466	2,380
Small/Medium Cities	217	373	316	534
Slowly Gentrifying				
Large Cities	379	484	535	649
Small/Medium Cities	98	145	149	154
Rapidly Gentrifying				
Large Cities	527	823	931	1,731
Small/Medium Cities	119	228	167	380

tracts were in large cities. In addition, 79.5% of slowly gentrifying and 81.6% of rapidly gentrifying tracts were in large cities. In 1970, large cities contained 78.6% of all gentrifiable tracts which means that educational gentrification in the 1970s was slightly concentrated in large cities.

During the 1980s, there was a large increase in the amount of educational gentrification in small/medium cities. The number of tracts in small/medium cities experiencing educational gentrification increased by 72% when compared to the number from the 1970s. The number of educationally gentrifying tracts in large cities increased by 44%. The number of rapidly gentrifying tracts in small/medium cities increased by 92% when compared to the 1970s, while the number of rapidly gentrifying tracts in large cities increased by 56%. Both types of cities experienced much smaller increases in the number of slowly gentrifying tracts. Small/medium cities saw the number of slowly gentrifying tracts increase by 48% relative to the 1970s, while the large cities experienced a 28% increase in the number of slowly gentrifying tracts.

The total number of educationally gentrifying tracts increased by only 6% in the 1990s relative to the number in the 1980s. However, the two types of cities had very different experiences in the 1990s. The number of educationally gentrifying tracts in large cities increased by 12% in the 1990s relative to the 1980s. The number in small/medium cities declined by over 15%. However, the entire small/medium city decline was due to a large decrease in the number of rapidly gentrifying tracts in these cities. The number of rapidly gentrifying tracts decreased by almost 27% in small/medium cities in the 1990s relative to the 1980s. The number of slowly gentrifying tracts increased by 2.8%. For large cities, the number of rapidly gentrifying tracts increased by 13% and the number of slowly gentrifying tracts increased by 10.5%.

Both types of cities experienced a large increase in the number of educationally gentrifying tracts in the 2000s relative to the 1990s. Large cities saw the total number of educationally gentrifying tracts increase by 62%, while the number in small/medium cities increased by 69%. For both types of cities, however, the increase was much higher for rapidly gentrifying tracts than for slowly gentrifying tracts. Large cities saw an 86% increase in the number of rapidly gentrifying tracts relative to the 1990s, while the number of slowly gentrifying tracts increased by 21%. For small/medium cities, the number of rapidly gentrifying tracts increased by 128% relative to the 1990s and the number of slowly gentrifying tracts increased by only 3%.

The results regarding the percentage of gentrifiable tracts that educationally gentrified, contained in Table 5.6, reveal that the two types of cities were very similar in the 1970s. In large cities, 21.7% of gentrifiable tracts experienced educational gentrification, while 19.5% of the gentrifiable tracts in small/medium cities did. During the 1980s, small/medium cities experienced a larger increase than larger cities. The percentage of gentrifiable tracts that gentrified increased by 7.2 percentage points for small/medium cities and 3.4 percentage points for large cities. In small/medium

TABLE 5.6 Percentage of Gentrifiable Tracts that Educationally Gentrify by City Size

	Percentage of Gentrifiable Tracts that Educationally Gentrify			
Total Gentrifying	1970s	1980s	1990s	2000s
Large Cities	21.73%	25.14%	26.75%	41.67%
Small/Medium Cities	19.53%	26.76%	21.31%	36.60%
Slowly Gentrifying				
Large Cities	9.09%	9.31%	9.76%	11.36%
Small/Medium Cities	8.82%	10.40%	10.05%	10.56%
Rapidly Gentrifying				
Large Cities	12.64%	15.83%	16.99%	30.30%
Small/Medium Cities	10.71%	16.36%	11.26%	26.05%

cities, 26.8% of the gentrifiable tracts experienced educational gentrification in the 1980s, while 25.1% of the gentrifiable tracts in large cities did so.

During the 1990s, the percentage of gentrifiable tracts that gentrified by education increased by 1.6 percentage points for large cities and decreased by 5.5 percentage points for small/medium cities. Finally, in the 2000s, both types of cities experienced a very large increase in the percentage of gentrifiable tracts that experienced educational gentrification. Large cities experienced a 14.9 percentage point increase, while small/medium cities experienced a 15.3 percentage point increase. During the 2000s, 41.7% of the gentrifiable tracts in large cities and 36.6% of the gentrifiable tracts in small/medium cities experienced educational gentrification.

Table 5.7 turns to the question of which cities experienced the highest levels of educational gentrification for each decade. The results are presented for two measures of gentrification intensity: the percentage of central city tracts that educationally gentrified and the percentage of gentrifiable tracts that educationally gentrified. The results are reported for the overall level of gentrification as well as separately for slow and rapid gentrification.

During the 1970s, there were two cities in which more than 30% of their central city tracts educationally gentrified: Minneapolis (31.9%) and Arlington (30.5%). Minneapolis also had the highest percentage of central city tracts slowly gentrify (12.1%) and rapidly gentrify (19.8%). In addition to Minneapolis, there were three cities in which at least 10% of the central city slowly gentrified by education: Jersey City (11.9%), Arlington (11.9%), and Salt Lake City (11.5%). Arlington (18.6%) was the only city other than Minneapolis that had at least 15% of its central city rapidly gentrify by education.

Turning to the percentage of gentrifiable tracts that gentrified, there were five cities in the 1970s in which at least 50% of the gentrifiable tracts experienced educational gentrification. The most interesting case is Arlington in which 18 out of 19 gentrifiable tracts (94.7%) gentrified by education. The other four were: Madison

TABLE 5.7 Cities with the Highest Levels of Educational Gentrification for Each Decade

Panel A: Percentage of Central City That Gentrifies by Education

	1970s		1980s		1990s		2000s	
Overall	Minneapolis, MN	31.9%	Jersey City, NJ	44.8%	Denver, CO	30.28%	Washington, DC	46.63%
	Arlington, VA	30.5%	Buffalo, NY	31.3%	Portland, OR	29.66%	Bridgeport, CT	44.74%
	Salt Lake City, UT	23.1%	Salt Lake City, UT	28.3%	Salt Lake City, UT	28.30%	Atlanta, GA	42.64%
	Seattle, WA	20.3%	Seattle, WA	27.8%	Seattle, WA	27.07%	Salt Lake City, UT	41.51%
	Jersey City, NJ	19.4%	Boston, MA	27.8%	Dayton, OH	24.07%	St. Louis, MO	41.51%
Slow	Minneapolis, MN	12.07%	Providence, RI	12.8%	Youngstown, OH	16.1%	Bridgeport, CT	21.05%
	Jersey City, NJ	11.94%	Buffalo, NY	12.5%	St. Paul, MN	12.3%	St. Louis, MO	14.15%
	Arlington, VA	11.86%	Jersey City, NJ	11.9%	Tacoma, WA	11.6%	Youngstown, OH	12.90%
	Salt Lake City, UT	11.54%	Lincoln, NE	10.6%	Cleveland, OH	11.3%	Long Beach, CA	11.93%
	Flint, MI	9.76%	Boston, MA	10.6%	Honolulu, HI	11.2%	Tacoma, WA	11.63%
Rapid	Minneapolis, MN	19.8%	Jersey City, NJ	32.8%	Portland, OR	25.5%	Atlanta, GA	38.76%
	Arlington, VA	18.6%	Salt Lake City, UT	24.5%	Salt Lake City, UT	22.6%	Washington, DC	38.20%
	Madison, WI	14.8%	Seattle, WA	21.8%	Denver, CO	21.8%	Salt Lake City, UT	33.96%
	Boston, MA	13.3%	Buffalo, NY	18.8%	Dayton, OH	20.4%	Denver, CO	28.37%
	Washington, DC	12.3%	Arlington, VA	18.6%	Seattle, WA	19.5%	Jersey City, NJ	28.36%

(Continued)

TABLE 5.7 (Continued)

Panel B   Percentage of Gentrifiable Tracts that Income-Gentrify

	1970s	1980s	1990s	2000s
Total	Arlington, VA 94.7%	Seattle, WA 74.00%	San Francisco, CA 71.43%	St. Petersburg, FL 73.33%
	Madison, WI 68.8%	Arlington, VA 72.73%	Seattle, WA 70.59%	Arlington, VA 71.43%
	Seattle, WA 65.9%	Jersey City, NJ 66.67%	Portland, WA 59.72%	Portland, OR 70.49%
	Lincoln, NE 55.0%	Honolulu, HI 54.05%	Salt Lake City, UT 55.56%	Seattle, WA 70.00%
	Minneapolis, MN 50.0%	Evansville, IN 53.33%	Honolulu, HI 53.85%	Atlanta, GA 69.62%
			Albuquerque, NM 53.85%	
Slow	Arlington, VA 36.84%	Lincoln, NE 29.17%	Albuquerque, NM 30.8%	St. Petersburg, FL 26.67%
	Greensboro, NC 33.33%	Arlington, VA 22.73%	Honolulu, HI 28.2%	Long Beach, CA 26.00%
	Fort Lauderdale, FL 28.57%	Evansville, IN 20.00%	Youngstown, OH 25.0%	Pittsburgh, PA 24.53%
	Seattle, WA 26.83%	Anaheim, CA 20.00%	Virginia Beach, VA 25.0%	Fresno, CA 24.32%
	Lincoln, NE 25.00%	Providence, RI 19.23%	Madison, WI 23.8%	Corpus Christi, TX 22.73%
	San Francisco, CA 25.00%			
Rapid	Arlington, VA 57.9%	Seattle, WA 58.0%	San Francisco, CA 57.1%	Virginia Beach, VA 66.67%
	Madison, WI 56.3%	Arlington, VA 50.0%	Portland, OR 51.4%	Arlington, VA 64.29%
	Seattle, WA 39.0%	Jersey City, NJ 48.9%	Seattle, WA 51.0%	Atlanta, GA 63.29%
	Honolulu, HI 37.5%	Greensboro, NC 43.8%	Salt Lake City, UT 44.4%	Seattle, WA 62.50%
	Minneapolis, MN 31.1%	Salt Lake City, UT 43.3%	Baton Rouge, LA 37.0%	Portland, OR 62.30%

(68.8%), Seattle (65.9%), Lincoln (55%), and Minneapolis (50%). Arlington had the highest percentage of gentrifiable tracts that slowly gentrified (36.8%) and rapidly gentrified (57.9%). Greensboro (33.3%) was the only city other than Arlington in which at least 30% of the gentrifiable tracts gentrified by education, while Madison (56.3%) was the only other city in which at least 50% of the gentrifiable tracts rapidly gentrified. During the 1970s, 92 of the 100 cities had at least one tract experience educational gentrification. In addition, 88 of the cities had at least one tract slowly gentrify and 77 of the cities had at least one tract rapidly gentrify. Educational gentrification was far more widespread than income gentrification in the 1970s.

During the 1980s, there were once again two cities in which at least 30% of the central city tracts experienced some sort of educational gentrification: Jersey City (44.8%) and Buffalo (31.3%). There were six cities in which at least 10% of the central city tracts slowly gentrified: Providence (12.8%), Buffalo (12.5%), Jersey City (11.9%), Lincoln (10.6%), Boston (10.6%), and St. Louis (10.4%). Led by Jersey City (32.8%), there were three cities in which at least 20% of the central city tracts educationally gentrified. The other two were Salt Lake City (24.5%) and Seattle (21.8%). It is worth noting that Seattle, Jersey City, and Salt Lake City were among the five cities with the highest percentage of central city tracts that gentrified by education in both the 1970s and the 1980s. Jersey City was among the leaders for the percentage of central city tracts slowly gentrifying in both decades, while Arlington was among the leaders for the percentage rapidly gentrifying in both decades.

Turning to the results regarding the percentage of gentrifiable tracts that gentrified reveals that there were nine cities in which at least 50% of the gentrifiable tracts gentrified in some way and three cities in which at least 60% gentrified. The top three were: Seattle (74.0%), Arlington (72.7%), and Jersey City (66.7%). There were four cities in which at least 20% of the gentrifiable tracts slowly gentrified: Lincoln (29.2%), Arlington (22.7%), Anaheim (20.0%), and Evansville (20.0%). In addition, there were two cities in which at least 50% of the gentrifiable tracts rapidly gentrified: Seattle (58.0%) and Arlington (50.0%). Seattle and Arlington were among the five cities with the highest percentage of gentrifiable tracts that gentrified in both the 1970s and the 1980s. Arlington was among the leaders for the percentage that both slowly and rapidly gentrified in both decades, while Seattle was among the leaders for the percentage that rapidly gentrified in each decade. During the 1980s, 97 out of the 100 cities had at least one tract experience educational gentrification in some way. In addition, 90 of the cities had at least one tract slowly gentrify and 90 of the cities had at least one tract rapidly gentrify.

During the 1990s, there was only one city, Denver (30.3%), in which at least 30% of the central city tracts educationally gentrified. There were seven cities in which at least 10% of the central city tracts slowly gentrified: Youngstown (16.1%), St. Paul (12.4%), Tacoma (11.6%), Cleveland (11.3%), Honolulu (11.2%), Minneapolis (10.3%), and Buffalo (10.1%). There were four cities in which at least

20% of the central city tracts rapidly gentrified: Portland (25.5%), Salt Lake City (22.6%), Denver (21.8%), and Dayton (20.4%). Seattle and Salt Lake City have been in the top five cities for the percentage of central city tracts that educationally gentrified for the 1970s, 1980s, and 1990s.

The results for the percentage of gentrifiable tracts that gentrified by education reveal that there were two cities (San Francisco (71.4%) and Seattle (70.6%)) in which the percentage of gentrifiable tracts that gentrified by education exceeded 70% and another four cities whose percentage exceeded 50%: Portland (59.7%), Salt Lake City (55.6%), Honolulu (53.9%), and Albuquerque (53.9%). Albuquerque (30.8%) was the only city whose percentage of gentrifiable tracts that slowly gentrified exceeded 30%. There were seven other cities whose percentage was at least 20%: Honolulu (28.2%), Youngstown (25.0%), Virginia Beach (25%), Madison (23.8%), Corpus Christi (21.1%), Columbus, GA (20.0%), and Arlington (20%). There were three cities in which the percentage of gentrifiable tracts that rapidly gentrified exceeded 50%: San Francisco (57.1%), Portland (51.4%), and Seattle (51.0%). It is interesting to note that these cities are all on the West Coast of the United States. During the 1990s, 97 of the 100 cities had at least one tract gentrify by education. In addition, 92 of the cities had at least one city slowly gentrify and 90 of the cities had at least one city rapidly gentrify.

The previously documented surge in educational gentrification in the 2000s is reflected in the individual city results as well. There were five cities in which the percentage of central city tracts that experienced some sort of educational gentrification exceeded 40% and an additional six whose percentage exceeded 30%. The five cities whose percentage exceeded 40% were: Washington (46.6%), Bridgeport (44.7%), Atlanta (42.6%), Salt Lake City (41.5%), and St. Louis (41.5%). In the previous decades, the highest number of cities whose percentage exceeded 30% was two in both the 1970s and the 1980s. Bridgeport (21.1%) was the only city in which the percentage of central city tracts that slowly gentrified exceeded 20%, but there were four additional cities whose percentage exceeded 10%: St. Louis (14.2%), Youngstown (12.9%), Long Beach (11.9%), and Tacoma (11.6%). There were three cities in which the percentage of central city tracts that rapidly gentrified exceeded 30%: Atlanta (38.8%), Washington (38.2%), and Salt Lake City (34.0%). An additional 12 cities had more than 20% of their central city tracts gentrify by education. Thus, there were 15 cities where the percentage of central city tracts that rapidly gentrified by education exceeded 20% in the 2000s. In the previous decades, the highest number of cities that exceeded 20% was four. Clearly, there was a very large increase in the amount of educational gentrification in the 2000s relative to the earlier decades.

Turning to the results for the percentage of gentrifiable tracts that gentrified by education in the 2000s reveals that there were 21 cities in which at least 50% of the gentrifiable tracts experienced some sort of educational gentrification. In the previous decades, the highest number was nine in the 1980s. There were eight cities in which the percentage of gentrifiable tracts that slowly gentrified exceeded 20%.

In addition, there were ten cities in which the percentage of gentrifiable tracts that rapidly gentrified by education exceeded 50%. During the 2000s, all 100 cities in the sample had at least one tract gentrify by education in some way. In addition, 96 of the cities had at least one tract slowly gentrify and 99 of the cities had at least tract rapidly gentrify. The results for the individual cities reflect the large increase in the amount of educational gentrification in the 2000s relative to the other decades and, also, reveal that the breadth of the gentrification also increased in the 2000s.

Table 5.8 contains an overview of how the educational gentrification trends compare to the income gentrification trends addressed in Chapter 4. The first thing that is clear from the table is that neighborhoods are more likely to gentrify by education than by income. However, income gentrification increased at a much faster rate than educational gentrification. In the 1970s, there were slightly more than five times as many tracts that gentrified by education than gentrified by income. During the 1980s, the ratio fell to slightly more than 2.5 times more educationally gentrifying tracts and the gap fell further during the 1990s to less than 1.5 times more educationally gentrifying tracts. There was a slight increase in the ratio in the 2000s to 1.7 but it remained well below the ratios for the 1970s and the 1980s.

The disparity between the number of tracts that gentrified by education and the number that gentrified by income was greatest among rapidly gentrifying tracts. In the 1970s, there were more than 14 times as many tracts that rapidly gentrified by education than by income. The ratio of the number of tracts that rapidly gentrified by education to the number of tracts that rapidly gentrified by income fell dramatically in the 1980s to 3.7, increased slightly to 4.3 in the 1990s, and fell again to 2.6 in the 2000s. For the sake of comparison, the ratio for slowly gentrifying tracts was 2.7 in the 1970s, fell to 1.8 in the 1980s, decreased further to 0.7 in the 1990s, and increased slightly to 0.9 in 2000s. Thus, in both the 1990s and 2000s, there were more tracts that slowly gentrified by income than slowly gentrified by education. The trends in Table 5.8 reflect the rapid increase in the amount of income gentrification in the 1980s and the 1990s that was documented in Chapter 4.

Table 5.9 compares the cities with the highest levels of educational and income gentrification in each decade. During the 1970s, there is no overlap between the five cities with the highest percentage of central city tracts that income gentrified and the five with the highest percentage that educationally gentrified. During the 1980s, only Jersey City appears on both lists. In the 1990s, there is once again no overlap. Finally, during the 2000s, there are three cities (Atlanta, Washington, and St. Louis) that appear on both lists. Thus, it appears that in the 2000s, there was more overlap between the areas that were gentrifying by income and those that were gentrifying by education.

When the analysis is limited to slowly gentrifying tracts, there are, once again, only two decades in which the lists overlap. Jersey City is on both lists in the 1980s and St. Louis is on both lists in the 2000. For rapidly gentrifying tracts, the only decade in which the two lists overlap is the 2000s when both Atlanta and Denver are on both lists.

TABLE 5.8 Comparison of Educational Gentrification Levels to Income Gentrification Levels

Total Number of Tracts Gentrifying by Education

	1970s	1980s	1990s	2000s
Total Gentrifying				
All Gentrifiable Tracts	1,123	1,680	1,782	2,914
Low-Income Gentrifiable	1,015	1,308	1,377	2,359
Very Low-Income Gentrifiable	108	372	405	555
Slowly Gentrifying				
All Gentrifiable Tracts	477	629	684	803
Low-Income Gentrifiable	430	512	514	649
Very Low-Income Gentrifiable	47	117	170	154
Rapidly Gentrifying				
All Gentrifiable Tracts	646	1,051	1,098	2,111
Low-Income Gentrifiable	585	796	863	1,710
Very Low-Income Gentrifiable	61	255	235	401

Total Number of Tracts Gentrifying by Income

	1970s	1980s	1990s	2000s
Total Gentrifying				
All Gentrifiable Tracts	222	644	1,221	1,691
Low-Income Gentrifiable	208	356	944	1,412
Very Low-Income Gentrifiable	14	288	277	279
Slowly Gentrifying				
All Gentrifiable Tracts	178	357	963	875
Low-Income Gentrifiable	171	305	744	751
Very Low-Income Gentrifiable	7	52	219	124
Rapidly Gentrifying				
All Gentrifiable Tracts	44	287	258	816
Low-Income Gentrifiable	37	51	200	661
Very Low-Income Gentrifiable	7	236	58	155

Ratio of Tracts that Gentrify by Education to Tracts That Gentrify by Income

	1970s	1980s	1990s	2000s
Total Gentrifying				
All Gentrifiable Tracts	5.06	2.61	1.46	1.72
Low-Income Gentrifiable	4.88	3.67	1.46	1.67
Very Low-Income Gentrifiable	7.71	1.29	1.46	1.99
Slowly Gentrifying				
All Gentrifiable Tracts	2.68	1.76	0.71	0.92
Low-Income Gentrifiable	2.51	1.68	0.69	0.86
Very Low-Income Gentrifiable	6.71	2.25	0.78	1.24
Rapidly Gentrifying				
All Gentrifiable Tracts	14.68	3.66	4.26	2.59
Low-Income Gentrifiable	15.81	15.61	4.32	2.59
Very Low-Income Gentrifiable	8.71	1.08	4.05	2.59

TABLE 5.9 Cities with the Highest Levels of Income and Educational Gentrification by Decade

Percentage of Central City Tracts that Gentrified by Income (Total)

1970s		1980s		1990s		2000s	
Des Moines, IA	10.5%	Jersey City, NJ	19.4%	Shreveport, LA	24.56%	Atlanta, GA	28.68%
Albuquerque, NM	9.1%	Madison, WI	16.4%	Grand Rapids, MI	20.00%	St. Louis, MO	26.42%
Santa Ana, CA	5.8%	Lincoln, NE	15.2%	Santa Ana, CA	19.23%	Flint, MI	24.39%
Shreveport, LA	5.3%	Fort Worth, TX	13.9%	Denver, CO	17.61%	Washington, DC	23.60%
Honolulu, HI	5.2%	Arlington, VA	13.6%	Miami, FL	17.17%	Denver, CO	23.40%

Percentage of Central City Tracts that Gentrified by Education (Total)

1970s		1980s		1990s		2000s	
Minneapolis, MN	31.90%	Jersey City, NJ	44.78%	Denver, CO	30.28%	Washington, DC	46.63%
Arlington, VA	30.51%	Buffalo, NY	31.25%	Portland, OR	29.66%	Bridgeport, CT	44.74%
Salt Lake City, UT	23.08%	Salt Lake City, UT	28.30%	Salt Lake City, UT	28.30%	Atlanta, GA	42.64%
Seattle, WA	20.30%	Seattle, WA	27.82%	Seattle, WA	27.07%	Salt Lake City, UT	41.51%
Jersey City, NJ	19.40%	Boston, MA	27.78%	Dayton, OH	24.07%	St. Louis, MO	41.51%
Cities on Both Lists	None	Jersey City		None		Atlanta	
						St. Louis	
						Washington	

Percentage of Central City Tracts that Gentrified by Income (Slow)

1970s		1980s		1990s		2000s	
Des Moines, IA	10.5%	Arlington, VA	13.6%	Grand Rapids, MI	18.0%	St. Louis, MO	16.04%
Albuquerque, NM	9.1%	Jersey City, NJ	13.4%	Santa Ana, CA	17.3%	Washington, DC	13.48%
Santa Ana, CA	5.8%	Paterson, NJ	9.1%	Spokane, WA	14.5%	Portland, OR	12.41%
New Orleans, LA	4.5%	Salt Lake City, UT	7.5%	Denver, CO	14.1%	Sacramento, CA	11.76%
Chicago, IL	3.8%	Norfolk, VA	6.3%	Shreveport, LA	14.0%	Newark, NJ	11.49%

(Continued)

TABLE 5.9 (Continued)

Percentage of Central City Tracts that Gentrified by Education (Slow)

1970s		1980s		1990s		2000s	
Minneapolis, MN	12.07%	Providence, RI	12.82%	Youngstown, OH	16.13%	Bridgeport, CT	21.05%
Jersey City, NJ	11.94%	Buffalo, NY	12.50%	St. Paul, MN	12.35%	St. Louis, MO	14.15%
Arlington, VA	11.86%	Jersey City, NJ	11.94%	Tacoma, WA	11.63%	Youngstown, OH	12.90%
Salt Lake City, UT	11.54%	Lincoln, NE	10.61%	Cleveland, OH	11.30%	Long Beach, CA	11.93%
Flint, MI	9.76%	Boston, MA	10.56%	Honolulu, HI	11.22%	Tacoma, WA	11.63%
Cities on Both Lists	None	Jersey City		None		St. Louis	

Percentage of Central City Tracts that Gentrified by Income (Rapid)

1970s		1980s		1990s		2000s	
Shreveport, LA	5.3%	Madison, WI	14.8%	Shreveport, LA	10.5%	Atlanta, GA	23.26%
Houston, TX	2.7%	Lincoln, NE	13.6%	Miami, FL	5.1%	Denver, CO	16.31%
Denver, CO	2.2%	Fort Worth, TX	11.9%	Flint, MI	4.9%	Cleveland, OH	12.99%
Fort Lauderdale, FL	2.1%	Columbus, OH	9.1%	Warren, MI	4.9%	Austin, TX	10.99%
Honolulu, HI	2.1%	Kansas City, MO	8.8%	Jersey City, NJ	4.5%	St. Louis, MO	10.38%

Percentage of Central City Tracts that Gentrified by Education (Rapid)

1970s		1980s		1990s		2000s	
Minneapolis, MN	19.83%	Jersey City, NJ	32.84%	Portland, OR	25.52%	Atlanta, GA	38.76%
Arlington, VA	18.64%	Salt Lake City, UT	24.53%	Salt Lake City, UT	22.64%	Washington, DC	38.20%
Madison, WI	14.75%	Seattle, WA	21.80%	Denver, CO	21.83%	Salt Lake City, UT	33.96%
Boston, MA	13.33%	Buffalo, NY	18.75%	Dayton, OH	20.37%	Denver, CO	28.37%
Washington, DC	12.29%	Arlington, VA	18.64%	Seattle, WA	19.55%	Jersey City, NJ	28.36%
Cities on Both Lists	None	None		None		Atlanta	
						Denver	

TABLE 5.10 Average Combined Ranking for Income and Educational Gentrification

Average Ranking for Percentage of Central City Tracts the Rapidly Gentrified by Income and the Percentage that Rapidly Gentrified by Education

1970s		1980s		1990s		2000s	
City	*Avg. Rank*	*City*	*Avg. Rank*	*City*	*Avg. Rank*	*City*	*Avg. Rank*
Washington, DC	9	Madison, WI	4	Denver, CO	7	Atlanta, GA	1
Denver, CO	9.5	Jersey City, NJ	5	Jersey City, NJ	8	Denver, CO	3
San Francisco, CA	13	Salt Lake City, UT	6	Atlanta, GA	8.5	Washington, DC	4
Boston, MA	15	Tampa, FL	15	San Francisco, CA	10	St. Louis, MO	6.5
Houston, TX	15.5	Kansas City, MO	15.5	Chicago, IL	10	Jersey City, NJ	8.5
Oklahoma City, OK	16	Portland, OR	17.5	Richmond, VA	11.5	Miami, FL	10
Chicago, IL	17	Norfolk, VA	17.5	Tulsa, OK	15.5	Salt Lake City, UT	11.5
Oakland, CA	17	Columbus, OH	17.5	Miami, FL	16	Austin, TX	12
Atlanta, GA	19.5	Evansville, IN	18.5	Dayton, OH	17.5	Chicago, IL	13.5
Philadelphia, PA	21	Greensboro, NC	19.5	Shreveport, LA	17.5	St. Paul, MN	14
Number of Cities with No Gentrification	22		6		9		1

Table 5.10 summarizes the cities that were most affected by the combination of income and educational gentrification. To assess this, a very simple measure is constructed in which the percentage of central city tracts that rapidly gentrify by income and the percentage that rapidly gentrify by education are calculated. The cities are then ranked from 1 to 100 for both income and educational gentrification. Cities that did not have any tracts gentrify during a decade are not given a ranking. Then the average ranking for each city is calculated and the cities with the lowest average ranking are identified as the cities most affected by the combination of income and educational gentrification. The top ten cities for each decade, their average ranking, and the number of cities with no tracts that gentrified by either income or education are reported.

During the 1970s, Washington, DC, and Denver were the only two cities with an average ranking below 10. Washington had the lowest average ranking of 9.0, while Denver had an average ranking of 9.5. Rounding out the top five were San Francisco (13.0), Boston (15.5), and Houston (15.5). There were 22 cities that had no tracts rapidly gentrify by income or by education in the 1970s.

In the 1980s, there were three cities with an average ranking below 10. Madison had the lowest average ranking of 4.0. Jersey City had an average ranking of 5.0, while Salt Lake City had an average ranking of 6.0. Tampa and Kansas City, both with an average ranking of 15.5, round out the top five. The number of cities in which there were no tracts that rapidly gentrified by either income or education fell from 22 in the 1970s to 6 in the 1980s.

In the 1990s, there were, once again, three cities with an average ranking below 10. Denver had the lowest average ranking at 7.0. Jersey City (8.0) and Atlanta (8.5) were the other two cities with average rankings below 10. The other two cities in the top five were San Francisco and Chicago who both had an average ranking of 10.0. The number of cities with zero tracts that rapidly gentrified by either income or education increased from 6 in the 1980s to 9 in the 1990s.

Finally, the 2000s saw the two lowest average rankings for any decade. Atlanta had the highest percentage of central city tracts that rapidly gentrified by both income and education and, therefore, had an average ranking of 1.0. There were four other cities with average rankings below 10. Denver (3.0), Washington (4.0), St. Louis (6.5), and Jersey City (8.5) were the other four cities. The number of cities in which no tracts rapidly gentrified by either income or education fell to 1 in the 2000s. This again emphasizes the extent to which gentrification was more widespread in the 2000s than in previous decades.

Table 5.11 turns to the question of which cities were most consistently impacted by income and educational gentrification over the entire timeframe included in this study. The table contains the cities with the lowest average rankings across all four decades for income-alone, education-alone, and the combined average for both income and education. As was mentioned above, cities that did not have any tracts gentrify during a decade were not given a ranking. This means that the only cities eligible to be among the leaders in Table 5.11 are those that had at least one tract gentrify in every decade.

TABLE 5.11 Average Ranking Across All Decades for Income, Education, and Combined

Income		Education		Combined	
City	Average Ranking	City	Average Ranking	City	Average Ranking
Denver, CO	12.5	Denver, CO	8.0	Denver, CO	10.3
Houston, TX	17.5	Boston, MA	8.5	Atlanta, GA	17.6
Chicago, IL	19.3	Portland, OR	13.5	Chicago, IL	20.5
Atlanta, GA	20.5	St. Paul, MN	14.3	Portland, OR	24.4
Memphis, TN	24.5	Atlanta, GA	14.8	Austin, TX	25.8
Detroit, MI	25.8	Jersey City, NJ	15.3	Boston, MA	26.0
Birmingham, AL	26.5	Sacramento, CA	19.8	Baltimore, MD	29.3
Shreveport, LA	26.5	New Orleans, LA	20.5	San Francisco, CA	29.9
Akron, OH	26.8	Minneapolis, MN	20.8	Jersey City, NJ	30.6
Kansas, City, KS	27.8	Chicago, IL	21.8	New Orleans, LA	30.8

There were two cities with an average ranking for income gentrification that was below 20. Denver had the lowest average ranking at 12.5 and Chicago was the other city with an average ranking of 19.5. There were seven cities with average rankings for educational gentrification below 10. Salt Lake City had the lowest average ranking at 3.5. Minneapolis had the second-lowest average ranking at 7.25 and was followed closely by Seattle at 7.75. The other four cities with average rankings below 10 were Boston (8.5), Denver (8.5), Jersey City (9.25), and Portland (9.75). Denver is the only city on both lists and, therefore, is the city with the lowest average combined ranking. In fact, Denver is the only city with an average combined ranking below 20. Denver had an average combined ranking of 10.5, while Chicago had the second-lowest average combined ranking at 20.875. Honolulu was the only other city with an average combined ranking below 30 at 26.25. Thus, the evidence points to Denver as the city most impacted by income and educational gentrification over the four decades running from 1970 to 2010.

Summary

This chapter has analyzed educational gentrification in U.S. cities from 1970 to 2010. The chapter initially sought to measure the amount of educational gentrification overall in the sample of U.S. cities. Educationally gentrifying tracts were identified in two ways. A tract was classified as *slowly* gentrifying if it was gentrifiable (using the definition described in Chapter 4) at the beginning of a decade and experienced a change in the percentage of residents aged 25 and over with at least a bachelor's degree during the decade that exceeded the change for the metropolitan area containing the city. A tract was defined to be *rapidly* gentrifying if its change exceeded the change in the metropolitan area by at least 50%.

The primary findings from the chapter are:

- Educational gentrification was more common than income gentrification in every decade. However, the gap between the number of tracts that gentrified by education and the number that gentrified by income decreased over time. The gap reached its lowest level in the 1990s and increased slightly in the 2000s.
- The number of tracts experiencing educational gentrification grew over time with each subsequent decade having more tracts that gentrified than the previous decade. This was true for both slow and rapid gentrification.
- The percentage of gentrifiable tracts that gentrified by education increased each decade. The largest increase was in the 2000s relative to the 1990s.
- Denver was the city most significantly affected by the combination of income and educational gentrification between 1970 and 2010.

Bibliography

Rucks-Ahidiana, Zawadi (2020). "Racial Composition and Trajectories of Gentrification in the United States", *Urban Studies*, 587(13): 2721–2741.

6

OCCUPATIONAL GENTRIFICATION IN U.S. CITIES 1970–2010

Introduction

The previous two chapters cataloged the trends in income and educational gentrification in U.S. cities from 1970 to 2010. While rising neighborhood incomes due to the in-migration of higher-income households and rising neighborhood education levels due to the in-migration of residents with college degree are the most common images of gentrifying neighborhoods, gentrification is also associated with the *professionalization* (Atkinson 2000) of gentrifiable neighborhoods. This chapter analyzes trends in occupational gentrification during the same period studied in the previous two chapters.

Occupational gentrification is identified by focusing on the change in the percentage of employed residents who work in two occupations identified by the Neighborhood Change Database: Professional and Technical Occupations and Executives, Managers, and Administrators. Occupational gentrification is identified in the same manner that both income and educational gentrification were identified in the previous two chapters. Tracts must first be gentrifiable (low-income, central city tracts) and experience an increase in the percentage of employed residents who work in these two occupations that exceed the change in the metropolitan area containing the central city in which the tract is located. Once again, tracts will be classified as slowly gentrifying if they simply have an increase in the percentage of residents employed in these two occupations that exceed the change at the metropolitan level. Tracts will be classified as undergoing rapid occupational gentrification if they experienced an increase in the percentage of workers in these two professional occupations that exceeds the change at the metropolitan level by at least 50%. The results are presented for the number of tracts that experienced either type of gentrification, those that only slowly gentrified and those that rapidly gentrified.

DOI: 10.1201/9781003217459-6

TABLE 6.1 Number of Occupationally-Gentrifying Tracts by Decade

	Number of Tracts that Occupationally-Gentrify			
Total Gentrifying	1970s	1980s	1990s	2000s
All Gentrifiable Tracts	2,460	2,572	2,589	3,459
Low-Income Gentrifiable	2,121	1,942	1843	2,797
Very Low-Income Gentrifiable	339	630	746	662
Slowly Gentrifying				
All Gentrifiable Tracts	586	901	654	257
Low-Income Gentrifiable	507	704	479	217
Very Low-Income Gentrifiable	79	197	175	40
Rapidly Gentrifying				
All Gentrifiable Tracts	1,874	1,671	1,935	3,202
Low-Income Gentrifiable	1,614	1,238	1,364	2,580
Very Low-Income Gentrifiable	260	433	571	622

Table 6.1 presents the results regarding the number of tracts that experience occupational gentrification for each decade from 1970 to 2010. One thing that immediately stands out is that the number of tracts gentrifying by occupation is much higher than the number that gentrified by income and education. This is especially true in the 1970s. A second difference is that the percentage of tracts that rapidly gentrified is also much higher than it was for the other two types of gentrification. Finally, the consistent upward trend in the number of gentrifying tracts that was present for both income and educational gentrification is not as evident for occupational gentrification.

During the 1970s, a total of 2,460 tracts experienced some sort of occupational gentrification. Of these, 76% of the tracts rapidly gentrified and 86% of the gentrifying tracts were in low-income gentrifiable neighborhoods rather than very low-income gentrifiable neighborhoods. During the 1980s, the percentage of tracts that occupationally gentrified increased by only 4.6%. This is much smaller than the increase in both income and educational gentrification from the 1970s to the 1980s. For both income and educational gentrification, there was a surge in gentrification activity in the 1980s compared to the 1970s. However, while the overall level of occupationally gentrifying tracts was essentially constant from the 1970s to the 1980s, there is a significant difference in the type of tracts that were gentrifying. The number of low-income gentrifiable tracts that gentrified decreased by 8%, while the number of very low-income tracts that experienced occupational gentrification increased by 86%. Because of this, the percentage of occupationally gentrifying tracts that were in low-income gentrifiable neighborhoods decreased from 86% in the 1970s to 76% in the 1980s. This trend is consistent with the trends for income and educational gentrification, both of which saw a surge in the number of very low-income tracts that gentrified in the 1980s relative to the number in the 1970s. The 1980s also saw a decline in the percentage of gentrifying tracts that rapidly gentrified relative to the 1970s. The percentage of rapidly gentrifying tracts fell from 76% in the 1970s to 65% in the 1980s. Thus, occupational gentrification was characterized by less dramatic changes in the 1980s than in the 1970s and a shift toward lower income gentrifiable neighborhoods.

The overall number of occupationally gentrifying tracts was almost constant from the 1980s to the 1990s. The total number of gentrifying tracts only increased by 0.7%. However, while the total number of gentrifying tracts remained almost constant, there was a 27% decrease in the number of slowly gentrifying tracts and a 16% increase in the number of rapidly gentrification. The percentage of gentrifying tracts that rapidly gentrified increased from 65% in the 1980s to 75% in the 1990s. In addition, there was a 5% decrease in the number of gentrifying tracts located in low-income gentrifiable tracts and an 18% increase in the number of gentrifying tracts located in very low-income gentrifiable tracts. The percentage of gentrifying tracts located in very low-income gentrifiable tracts increased from 24% in the 1980s to 29% in the 1990s. The 1990s were characterized by a decrease in slowly gentrifying tracts located in low-income gentrifiable neighborhoods and an increase in rapidly gentrifying tracts located in very low-income neighborhoods.

Like both income and educational gentrification, there was a surge in occupational gentrification during the 2000s. The total number of gentrifying tracts increased by 34% relative to the 1990s. The number of slowly gentrifying tracts decreased by 61% relative to the 1990s and the number of rapidly gentrifying tracts increased by 66%. The percentage of gentrifying tracts that rapidly gentrified increased from 75% in the 1990s to 93% in the 2000s. Thus, the surge in occupational gentrification in the 2000s was mainly due to a large increase in the number of rapidly gentrifying neighborhoods. In addition, there was a 51% increase in the number of gentrifying tracts located in low-income gentrifiable neighborhoods and an 11% decrease in the number located in very low-income gentrifiable neighborhoods. The percentage of gentrifying tracts located in very low-income gentrifiable tracts decreased from 29% in the 1990s to 19% in the 2000s. The decline in slowly gentrifying tracts occurred in both low-income and very low-income gentrifiable tracts. The number of slowly gentrifying tracts located in low-income gentrifiable tracts decreased by 55% and the number located in very low-income gentrifiable tracts decreased by 77%. However, the increase in rapidly gentrifying tracts took place almost entirely within low-income gentrifiable tracts. There was an 89% increase in the number of rapidly gentrifying tracts located in low-income gentrifiable tracts and a 9% increase in the number located in very low-income gentrifiable tracts. The percentage of rapidly gentrifying tracts located in very low-income gentrifiable tracts decreased from 30% in the 1990s to 19% in the 2000s. The surge in occupational gentrification in the 2000s was primarily due to a large increase in rapidly gentrifying tracts located in low-income gentrifiable neighborhoods.

TABLE 6.2 Percentage of Gentrifiable Tracts that Occupationally Gentrify by Decade

	Percentage of Gentrifiable Tracts that Occupationally Gentrify			
Total Gentrifying	1970s	1980s	1990s	2000s
All Gentrifiable Tracts	46.59%	39.01%	37.18%	48.14%
Low-Income Gentrifiable	45.93%	38.09%	35.29%	47.86%
Very Low-Income Gentrifiable	51.21%	42.17%	42.87%	49.37%
Slowly Gentrifying				
All Gentrifiable Tracts	11.10%	13.67%	9.39%	3.58%
Low-Income Gentrifiable	10.98%	13.81%	9.17%	3.71%
Very Low-Income Gentrifiable	11.93%	13.19%	10.06%	2.98%
Rapidly Gentrifying				
All Gentrifiable Tracts	35.49%	25.35%	27.79%	44.57%
Low-Income Gentrifiable	34.95%	24.28%	26.12%	44.15%
Very Low-Income Gentrifiable	39.27%	28.98%	32.82%	46.38%

Table 6.2 measures the amount of occupational gentrification as the percentage of gentrifiable tracts that gentrify during each decade. The trends in Table 6.2 are very different than those reported for income and educational gentrification. Both income and educational gentrification experienced a consistent increase in the percentage of gentrifiable tracts that gentrified from decade to decade. For occupational gentrification, however, the percentage is very high in the 1970s relative to the percentages for income and educational gentrification and then decreases for the next two decades. Then, as was the case for income and educational gentrification, there is a large increase in the percentage of gentrifiable tracts that occupationally gentrify in the 2000s relative to the 1990s.

In the 1970s, almost half (46.6%) of all gentrifiable tracts experienced occupational gentrification. The percentage fell to 39.0% in the 1980s and fell further to 37.2% in the 1990s. Then, in the 2000s, the percentage reached its highest level of 48.2%. Thus, the 2000s were characterized by a large increase in the percentage of gentrifiable tracts that experienced all three types of gentrification.

The decline in the percentage of gentrifiable tracts that experienced occupational gentrification in the 1980s relative to the 1970s was entirely due to a large decrease in the percentage of gentrifiable tracts that rapidly gentrified. The percentage of gentrifiable tracts that rapidly gentrified decreased from 35.5% in the 1970s to 25.4% in the 1980s, while the percentage of gentrifiable tracts that slowly gentrified increased from 11.1% in the 1970s to 13.7% in the 1980s. In addition, the decline in the percentage of gentrifiable tracts that rapidly gentrified was experienced in both low-income and very low-income gentrifiable neighborhoods to almost the same extent. Both types of gentrifiable neighborhoods saw their percentage that experienced rapid occupational gentrification fall by approximately 10 percentage points. Regarding slow occupational gentrification, in low-income gentrifiable neighborhoods, the percentage that slowly gentrified increased by 2.8 percentage points compared to 1.3 percentage points in very low-income neighborhoods.

In the 1990s, the percentage of gentrifiable tracts that occupationally gentrified once again decreased relative to the 1980s. However, the decrease of 1.8 percentage points was much smaller than the 7.6-percentage point decline in the 1980s relative to the 1970s. This time, however, the decline can be attributed entirely to a decrease in the probability that gentrifiable tracts slowly gentrified. The probability that a gentrifiable tract slowly gentrified by occupation decreased from 13.7% in the 1980s to 9.4% in the 1990s. The decline in the probability that a gentrifiable tracts slowly gentrified was experienced almost equally by low-income and very low-income gentrifiable tracts. For low-income gentrifiable tracts, the probability that they slowly gentrified with respect to occupation decreased from 13.8% in the 1980s to 9.2% in the 1990s, a decrease of 4.6 percentage points. Very low-income gentrifiable tracts experienced a decrease of 3.1 percentage points. At the same time, however, the percentage of gentrifiable tracts that rapidly gentrified increased from 25.4% in the 1980s to 27.8% in the 1990s. Both low-income and very low-income gentrifiable tracts experienced an increase in the probability that they rapidly gentrified. For low-income gentrifiable tracts, the increase was 1.8 percentage points while very low-income gentrifiable tracts experienced an increase of 3.8 percentage points.

Finally, as was mentioned above, the probability that gentrifiable tracts occupationally gentrified in the 2000s was much higher than the probability in the 1990s. The probability increased from 37.2% in the 1990s to 48.1% in the 2000s, an increase of 9.9 percentage points. However, the increase was entirely driven by a very large increase in the probability that gentrifiable tracts rapidly gentrified with respect to occupation. The probability that a gentrifiable tract rapidly gentrified with respect to occupation increased from 27.8% in the 1990s to 44.6% in the 2000s, an increase of 16.8 percentage points. The large increase in the probability that a gentrifiable tract rapidly gentrified with respect to occupation was experienced by both low-income and very low-income gentrifiable tracts. The probability that low-income gentrifiable tracts rapidly gentrified with respect to occupation increased from 26.1% in the 1990s to 44.2% in the 2000s, an increase of 18.0 percentage points while the probability that very low-income gentrifiable tracts rapidly gentrified with respect to occupation increased from 32.8% in the 1990s to 46.4% in the 2000s, an increase of 13.6 percentage points.

At the same time, the probability that a gentrifiable tract slowly gentrified with respect to occupation decreased from 9.4% in the 1990s to 3.6% in the 2000s, a decline of 5.8 percentage points. The decline was experienced in both low-income and very low-income gentrifiable tracts. For low-income tracts, that decline was 5.5 percentage points, while in very low-income tracts, the probability that a tract slowly gentrified by occupation fell by 7.1 percentage points.

TABLE 6.3 Number of Occupationally Gentrifying Tracts by Census Region

	Number of Occupationally Gentrifying Tracts			
Total Gentrifying	1970s	1980s	1990s	2000s
Northeast	552	845	688	963
Midwest	1,248	698	696	816
South	651	576	700	890
West	561	453	505	790
Slowly Gentrifying				
Northeast	164	279	178	66
Midwest	347	270	181	87
South	156	223	161	58
West	83	129	134	46
Rapidly Gentrifying				
Northeast	388	566	510	897
Midwest	901	428	515	729
South	495	353	539	832
West	478	324	371	744

Table 6.3 provides the regional breakdown for occupationally gentrifying tracts. During the 1970s, occupational gentrification was most concentrated in the Midwest. The Midwest contained 28% of the gentrifiable tracts in 1970 and 41% of the tracts that occupationally gentrified. Occupational gentrification in the 1970s was least concentrated in the Northeast which had 26% of the gentrifiable tracts in 1970 and only 18% of the tracts that occupationally gentrified.

During the 1980s, the number of tracts that gentrified by occupation decreased in every region except for the Northeast. In the Northeast, the number of occupationally gentrifying tracts increased by 53% compared to the 1970s. The biggest decrease relative to the 1970s was in the Midwest where the number of occupationally gentrifying tracts decreased by 44%. In the South, the number decreased by 12% while the number decreased by 19% in the West. The increase in the number of occupationally gentrifying tracts in the Northeast was such that the region went from being the region with the lowest concentration of occupationally gentrifying tracts to the region with the highest concentration. In 1980, 27% of the gentrifiable tracts were in the Northeast, while 33% of the occupationally gentrifying tracts in the 1980s were in the Northeast. The South replaced the Northeast as the region with the lowest concentration of occupationally gentrifying tracts with 27% of the gentrifiable tracts in 1980s and only 22% of the occupationally gentrifying tracts in the 1980s.

During the 1990s, the number of occupationally gentrifying tracts increased in the South and West while decreasing in the Northeast and Midwest. The South had the largest increase in the number of occupationally gentrifying tracts relative to the 1980s with a 21.5% increase while the number in the West increased by 11.5%. The biggest decrease in the number of occupationally gentrifying tracts was in the

Northeast where the number of occupationally gentrifying tracts fell by 18.6%. The number in the Midwest fell by 0.3%. Interestingly, occupational gentrification in the 1990s was distributed almost proportionally across the regions. The Northeast remained the region with the highest concentration with 25.7% of the gentrifiable tracts in 1990 and 26.5% of the occupationally gentrifying tracts in the 1990s. The lowest concentration was in the Midwest which had 27.8% of the gentrifiable tracts in 1990 and 26.9% of the occupationally gentrifying tracts in the 1990s.

Finally, in the 2000s, the number of occupationally gentrifying tracts increased in all four regions relative to the 1990s. The increase ranged from 17.2% in the Midwest to 56.4% in the West. The increase in the West was such that the region now had the highest concentration of occupationally gentrifying tracts with 20% of the gentrifiable tracts in 2000 and 23% of the occupationally gentrifying tracts in the 2000s. The lowest concentration was still in the Midwest which had 26% of the gentrifiable tracts in 2000 and only 24% of the occupationally gentrifying tracts in the 2000s. However, like 1990s, the distribution of occupational gentrification across the regions was much more proportional than it was in the 1970s and 1980s.

Table 6.4 provides the regional differences with respect to the percentage of gentrifiable tracts that occupationally gentrified during each decade. Recall that, since a tract must be gentrifiable before it can gentrify, simply counting the number of tracts that gentrify may not provide a complete picture regarding the intensity of gentrification in an area. Analyzing the percentage of gentrifiable tracts that experience gentrification provides a more complete picture regarding which regions were experiencing the most gentrification.

TABLE 6.4 Percentage of Gentrifiable Tracts that Occupationally Gentrify by Census Region

	Percentage of Gentrifiable Tracts that Occupationally-Gentrify			
Total Gentrifying	1970s	1980s	1990s	2000s
Northeast	16.6%	27.4%	26.6%	43.9%
Midwest	21.2%	23.6%	21.5%	35.8%
South	22.3%	22.2%	23.7%	37.8%
West	25.9%	29.8%	32.6%	46.5%
Slowly Gentrifying				
Northeast	6.9%	11.8%	10.4%	13.7%
Midwest	8.1%	9.7%	10.1%	10.6%
South	9.1%	7.4%	8.5%	9.6%
West	12.9%	9.2%	10.6%	10.9%
Rapidly Gentrifying				
Northeast	9.7%	15.6%	16.3%	30.2%
Midwest	13.1%	13.9%	11.4%	25.2%
South	13.2%	14.9%	15.2%	28.3%
West	13.0%	20.6%	22.1%	35.6%

In the 1970s, the West had the highest percentage of gentrifiable tracts that experienced occupational gentrification with 25.9% of the gentrifiable tracts in the West gentrifying by occupation. The lowest percentage was in with Northeast at 16.6%.

The ranking remains the same when only tracts that slowly gentrified by occupation are included. The West had the highest percentage of tracts that slowly gentrified by occupation at 12.9%, while the Northeast had the lowest percentage at 6.9%. When focusing on tracts that rapidly gentrified by occupation, the Midwest, South, and West regions have virtually identical percentages, ranging from 13.0 to 13.2%, while the Northeast, once again, has the lowest percentage at 9.7%.

The percentages for overall occupational gentrification increased in the 1980s relative to the 1970s in every region except for the South where the percentage fell by 0.1 percentage points. The largest increase was in the Northeast where the percentage of gentrifiable tracts that gentrified by occupation increased from 16.6% to 27.4%, an increase of 10.8 percentage points. The West had the next largest increase with an increase of 3.9 percentage points, while the increase in the Midwest was 2.4 percentage points. The increase in the West was such that the region had the highest percentage of gentrifiable tracts experiencing some sort of occupational gentrification at 29.8%. The South had the lowest percentage in the 1980s at 22.2%.

Focusing only on slow gentrification reveals that the percentage of gentrifiable tracts that slowly gentrified by occupation increased in two regions (Northeast and Midwest) and fell in two regions (South and West). The largest increase was for the Northeast where the percentage of gentrifiable tracts that slowly gentrified by occupation increased from 6.9% in the 1970s to 11.8%, an increase of 4.9 percentage points. The largest decline was in the West where the percentage fell by 3.7 percentage points. The large increase in the Northeast meant that the region went from having the lowest percentage of tracts that slowly gentrified in the 1970s to having to highest percentage, 11.8%, in the 1980s.

The percentage of gentrifiable tracts that experienced rapid occupational gentrification increased in every region in the 1980s relative to the 1970s. The largest increase was in the West where the percentage increased by 7.6 percentage points. The second largest increase was 5.9 percentage points in the Northeast. The increase for the South was 1.7 percentage points, while the Midwest had the smallest increase of 0.8 percentage points. The West had the highest percentage of gentrifiable tracts that rapidly gentrified by occupation at 20.6% while the Midwest had the lowest percentage at 13.9%.

In the 1990s, the percentage of gentrifiable tracts that experienced either slow or rapid gentrification increased in two regions (South and West) and fell in the other two regions (Northeast and Midwest). The West, with an increase of 2.8 percentage points, had the largest increase while the Midwest had the largest decrease of 2.1 percentage points. The West remained the region with the highest percentage of gentrifiable tracts that experienced some sort of occupational gentrification at 32.6%. The decline in the percentage in the Midwest meant that the region had the lowest percentage in the 1990s at 21.5%.

The percentage of gentrifiable tracts that slowly gentrified in the 1990s increased in every region except for the Northeast where the percentage fell from 11.8% in the 1980s to 10.4% in the 1990s. The increases in the other three regions were small with the West having the largest increase of 1.4 percentage points. The West had the highest percentage of gentrifiable tracts that slowly gentrified by occupation in the 1990s at 10.6%, while the South had the lowest percentage at 8.5%.

The percentage of gentrifiable tracts that rapidly gentrified by occupation in the 1990s increased in every region except for the Midwest in the 1990s. As was the case for slow gentrification, the increases were small, with the West experiencing the largest increase of 1.5 percentage points. The decrease in the Midwest was 2.5 percentage points. The West once again had the highest percentage of gentrifiable tracts that rapidly gentrified by occupation at 22.1%, while the Midwest once again had the lowest percentage at 11.4%.

The results in Chapters 4 and 5, as well as the results from earlier in this chapter, indicated that there was a large increase in the amount of gentrification in U.S. cities in the 2000s relative to the previous decades. The results in Table 6.4 show that this was the case as well. All four regions experienced a double-digit increase in the percentage of gentrifiable tracts that occupationally gentrified in the 2000s relative to the 1990s. The Northeast had the largest increase of 17.3 percentage points while the second largest increase was in the Midwest where the percentage increased by 14.3 percentage points. The percentage in the South increased by 14.1 percentage points, while the West had the smallest increase of 13.9 percentage points. The highest percentage of gentrifiable tracts that occupationally gentrified was in the West at 46.5% while the lowest value of 35.8% was in the Midwest. However, all four regions had values that exceeded any value from the previous three decades. Once again, the results indicate that there was a large increase in the amount of occupational gentrification in the 2000s relative to the earlier decades.

Focusing on the results for slowly gentrifying tracts reveals that most of the increase in occupational gentrification in the 2000s was an increase in rapid gentrification. The percentage of gentrifiable tracts that slowly gentrified by occupation increased in all four regions. However, the magnitudes of the increases represent only a small portion of the total increases. The largest increase in slow gentrification in the 2000s was for the Northeast where the percentage increased by 3.3 percentage points while the smallest increase was for the West at 0.3 percentage points. The Northeast was the region with the highest percentage of gentrifiable tracts that slowly gentrified by occupation in the 2000s with a percentage of 13.7%, while the South had the lowest percentage of 9.6%.

All four regions experienced a large increase in the percentage of gentrifiable tracts that rapidly gentrified by occupation in the 2000s relative to the 1990s. The increases were remarkably similar for the regions. The largest increases were for the Northeast (13.9 percentage points) and the Midwest (13.8 percentage points) and the smallest increases were for the West (13.5 percentage points) and South (13.1 percentage points). The highest percentage of gentrifiable tracts that rapidly

TABLE 6.5 Number of Tracts that Occupationally Gentrify by City Size

	Number of Tracts			
	1970s	1980s	1990s	2000s
Total Gentrifying				
Large Cities	1,953	2,088	2,085	2,784
Small/Medium Cities	507	484	504	675
Slowly Gentrifying				
Large Cities	466	713	517	193
Small/Medium Cities	120	188	137	64
Rapidly Gentrifying				
Large Cities	1,487	1,375	1,568	2,591
Small/Medium Cities	387	296	367	611

gentrified by occupation in the 2000s was in the West at 35.6% while the lowest percentage was in the Midwest at 25.2%. All four regions had values in the 2000s that exceeded any of the values from the previous three decades. As was the case for both income and educational gentrification, there was a large increase in the amount of occupational gentrification in the 2000s.

Table 6.5 provides the results for the number of occupationally gentrifying tracts by city size. During the 1970s, the distribution of occupationally gentrifying tracts across the two city types is almost identical to the distribution of gentrifiable tracts. Large cities contained 79% of the gentrifiable tracts and 79% of the occupationally gentrifying tracts. The proportionality remains when the gentrifying tracts are split into slowly and rapidly gentrifying tracts.

During the 1980s, occupational gentrification increased slightly more in large cities than in small/medium cities. The number of occupationally gentrifying tracts in large cities increased by almost 7% when compared to the 1970s while the number in small/medium cities decreased by 4.5%. Both types of cities experienced similar increases in the number of slowly gentrifying cities (53% for large cities; 57% for small/medium cities) and both types of cities saw the number of rapidly gentrifying tracts decrease in the 1980s relative to the 1970s. However, the decrease in rapidly gentrifying tracts was much larger in small/medium cities (23.5%) than in large cities (7.5%).

During the 1990s, the number of occupationally gentrifying tracts in large cities decreased by 0.1% and, thus, was virtually identical to the number in the 1980s. For small/medium cities, the number of occupationally gentrifying tracts increased by 4.1%. Both types of cities saw comparable decreases in the number of slowly gentrifying tracts (27.5% for large cities and 27.1% for small/medium cities) and both types of cities experienced an increase in the number of rapidly gentrifying tracts. However, the increase in small/medium cities (24.0%) was larger than the increase in large cities (14.0%).

As would be expected from the previous results, both types of cities experienced a large increase in the number of occupationally gentrifying tracts in the

TABLE 6.6 Percentage of Gentrifiable Tracts that Occupationally Gentrify by City Size

	Percentage of Gentrifiable Tracts that Occupationally Gentrify			
Total Gentrifying	1970s	1980s	1990s	2000s
Large Cities	46.85%	40.16%	38.05%	48.74%
Small/Medium Cities	45.63%	34.72%	33.99%	46.26%
Slowly Gentrifying				
Large Cities	11.18%	13.71%	9.43%	3.38%
Small/Medium Cities	10.80%	13.49%	9.24%	4.39%
Rapidly Gentrifying				
Large Cities	35.67%	26.45%	28.61%	45.36%
Small/Medium Cities	34.83%	21.23%	24.75%	41.88%

2000s relative to the 1990s. Both types of cities experienced similar increases, with large cities experiencing a 33.5% increase and small/medium cities experiencing a 33.9% increase. Both types of cities also experienced a large decline in the number of slowly gentrifying tracts and a large increase in the number of rapidly gentrifying tracts. Large cities saw the number of slowly gentrifying tracts decrease by 62.7% while the number of rapidly gentrifying tracts increased by 65.2% relative to the 1990s. For small/medium cities, the number of slowly gentrifying tracts decreased by 53.3% and the number of rapidly gentrifying tracts increased by 66.5%.

Table 6.6 contains the results for the percentage of gentrifiable tracts that occupationally gentrify for both types of cities. In the 1970s, the probability that a gentrifiable tract experienced occupational gentrification was essentially the same for large and small/medium cities. In large cities, 46.9% of gentrifiable tracts gentrified, while 45.6% of gentrifiable tracts gentrified in small/medium cities. During the 1980s, the percentage fell for both types of cities but the decline was larger for small/medium cities than for large cities. The percentage for small/medium cities fell by almost 11 percentage points to 34.7%, while the percentage for large cities fell by 6.7 percentage points to 40.2%. The percentages for both types of cities continued to decline in the 1990s. The percentage for large cities fell by 2.1 percentage points to 38.1%, while the percentage for small/medium cities fell by 0.7 percentage points to 34.0%. Finally, both types of cities experienced large increases in the percentage of gentrifiable tracts that occupationally gentrified in the 2000s. The increase for small/medium cities was slightly larger than the increase for large cities. The percentage for small/medium cities increased by 12.3 percentage points to 46.3%, while the percentage for large cities increased by 10.7 percentage points to 48.7%. Both types of cities experienced the highest percentage of gentrifiable tracts that occupationally gentrified in the 2000s.

Table 6.7 switches the focus to the question of which cities were most impacted by occupational gentrification during each decade. Panel A of the table measures the impact of gentrification as the percentage of central city tracts that occupationally

TABLE 6.7 Cities with the Highest Levels of Occupational Gentrification for Each Decade

Panel A: Percentage of Central City That Gentrified by Occupation

	1970s	1980s	1990s	2000s
Overall	Minneapolis, MN 49.1% Hartford, CT 41.0% St. Louis, MO 38.7% Washington, DC 38.5% Oakland, CA 33.9%	Jersey City, NJ 44.8% Buffalo, NY 41.3% Bridgeport, CT 39.5% Minneapolis, MN 34.5% St. Louis, MO 32.1%	Newark, NJ 40.23% Portland, OR 34.48% Cleveland, OH 33.90% Richmond, VA 30.30% Dayton, OH 29.63%	Bridgeport, CT 63.16% Gary, IN 58.06% Washington, DC 48.31% Atlanta, GA 44.96% Tacoma, WA 44.19%
Slow	Hartford, CT 15.38% Baltimore, MD 11.00% Washington, DC 10.61% Newark, NJ 10.47% Gary, IN 9.68%	Buffalo, NY 22.5% Providence, RI 20.5% St. Louis, MO 14.2% St. Paul, MN 13.6% Salt Lake City, UT 13.2%	Bridgeport, CT 15.8% Miami, FL 12.1% Madison, WI 11.5% Milwaukee, WI 11.0% St. Louis, MO 9.4%	Tacoma, WA 11.63% Hartford, CT 10.26% Providence, RI 7.69% Baltimore, MD 7.00% Richmond, VA 6.06%
Rapid	Minneapolis, MN 44.0% St. Louis, MO 30.2% Washington, DC 27.9% St. Paul, MN 27.2% Salt Lake City, UT 26.9%	Jersey City NJ 37.3% Bridgeport, CT 26.3% Minneapolis, MN 24.1% Gary, IN 22.6% Washington, DC 22.5%	Newark, NJ 32.2% Cleveland, OH 26.3% Salt Lake City, UT 24.1% Portland, OR 22.6% Dayton, OH 22.5%	Bridgeport, CT 63.16% Gary, IN 58.06% Washington, DC 46.63% Atlanta, GA 44.19% Newark, NJ 40.23%

(*Continued*)

TABLE 6.7 (Continued)

Panel B: Percentage of Gentrifiable Tracts that Occupationally-Gentrify

	1970s		1980s		1990s		2000s	
Total	Seattle, WA	82.9%	Jersey City, NJ	66.67%	San Francisco, CA	82.14%	Fort Lauderdale, FL	83.33%
	Portland, OR	79.5%	Youngstown, OH	66.67%	Seattle, WA	72.55%	Portland, OR	80.33%
	Arlington, VA	78.9%	Gary, IN	64.29%	Portland, OR	69.44%	Atlanta, GA	73.42%
	Minneapolis, MN	77.0%	Arlington, VA	63.64%	El Paso, TX	66.67%	Washington, DC	71.07%
	Sacramento, CA	76.7%	Buffalo, NY	61.11%	Tulsa, OK	63.41%	Seattle, WA	70.00%
					Albuquerque, NM	53.85%		
Slow	Jacksonville, FL	23.81%	Buffalo, NY	33.33%	Madison, WI	33.3%	Tacoma, WA	17.86%
	Birmingham, AL	23.53%	Providence, RI	30.77%	Greensboro, NC	21.4%	Seattle, WA	12.50%
	Hartford, CT	19.35%	Knoxville, TN	27.27%	Milwaukee, WI	20.5%	Albuquerque, NM	12.50%
	Baltimore, MD	19.13%	Wichita, KS	26.67%	Lincoln, NE	20.0%	Tampa, FL	11.76%
	Omaha, NE	18.42%	Louisville, KY	25.00%	Springfield, MA	20.0%	Greensboro, NC	11.76%
Rapid	Seattle, WA	78.0%	Arlington, VA	78.0%	San Francisco, CA	59.1%	Fort Lauderdale, FL	83.33%
	Spokane, WA	75.0%	Jersey City, NJ	75.0%	El Paso, TX	55.6%	Atlanta, GA	72.15%
	Minneapolis, MN	68.9%	Youngstown, OH	68.9%	Seattle, WA	50.0%	Portland, OR	72.13%
	Madison, WI	68.8%	Gary, IN	68.8%	Tulsa, OK	50.0%	Washington, DC	68.60%
	San Jose, CA	68.2%	Long Beach, CA	68.2%	Portland, OR	45.1%	El Paso, TX	67.86%

gentrified, while Panel B measures the impact as the percentage of gentrifiable tracts that occupationally gentrified. The results in each panel are provided for the percentage of tracts that occupationally gentrified in any way and separately for slow and rapid occupational gentrification.

In the 1970s, according to panel A, there were two cities in which at least 40% of the central city occupationally gentrified. Minneapolis had the highest percentage at 49.1%, while 41.0% of the central city tracts in Hartford occupationally gentrified. There were also six cities (St. Louis, Washington, Oakland, St. Paul, Jersey City, and Salt Lake City) in which at least 30% of the central city experienced occupational gentrification. In total, 98 of the 100 cities had at least one central city tract occupationally gentrify.

Hartford had the highest percentage of central city tracts slowly gentrify at 15.4%. There were three other cities (Baltimore, Washington, and Newark) in which at least 10% of the central city slowly gentrified. In total, 82 of the 100 cities had at least one central city tract slowly gentrify during the 1970s. Minneapolis had the highest percentage of central city tracts rapidly gentrify during the 1970s at 44.0% and was the only city with at least 40% of the central city rapidly gentrifying by occupation. St. Louis had the second highest level of rapid gentrification during the 1970s at 30.2% and was the only other city with at least 30% of central city tracts experiencing rapid occupational gentrification. In total, 97 of the 100 cities had at least one central city tracts rapidly gentrify by occupational gentrification during the 1970s.

Turning now to the results in Panel B for the 1970s, among cities with at least ten gentrifiable tracts, Seattle, at 82.9%, had the highest percentage of gentrifiable tracts occupationally gentrify. Seattle was the only city with over 80% of its gentrifiable tracts experiencing occupational gentrification while there were seven cities in which at least 70% of the gentrifiable tracts occupationally gentrified.[1] During the 1970s, there were 47 cities in which at least 50% of their gentrifiable tracts experienced occupational gentrification. Jacksonville, at 23.8%, and Birmingham, at 23.5%, had the highest percentages of gentrifiable tracts that slowly gentrified in the 1970s and were the only cities in which at least 20% of their gentrifiable tracts occupationally gentrified. Seattle, at 78.0%, and Spokane, at 75.0%, had the highest percentage of gentrifiable tracts that experienced rapid occupational gentrification in the 1970s and were the only two cities in which at least 70% of the gentrifiable tracts experienced rapid occupational gentrification. There were an additional 12 cities in which at least 60% of the gentrifiable tracts rapidly gentrified by occupation and another 25 cities in which at least 50% of the gentrifiable tracts experienced rapid occupational gentrification.

Turning to the 1980s, according to Panel A, which measures the impact of gentrification as the percentage of central city tracts that occupationally gentrify, there were, once again, two cities in which at least 40% of the central city occupationally gentrified. Jersey City had the highest percentage at 44.8%, while 41.3% of the central city tracts in Buffalo occupationally gentrified. The level of occupational gentrification remained high in Minneapolis with 34.5% of the central city tracts occupationally gentrifying, the fourth highest level in the 1980s. In addition to Minneapolis, there were seven other cities (Bridgeport, St. Louis, Washington, Oakland, Providence, Salt Lake City, and Boston) in which at least 30% of the central city occupationally gentrified in the 1980s. Both Minneapolis and St. Louis were among the cities with the five highest levels of overall occupational gentrification in both the 1970s and 1980s. Minneapolis was 1st in the 1970s and 4th in the 1980s, while St. Louis was 3rd in the 1970s and 5th in the 1980s. In total, 98 of the 100 cities had at least one central city tract occupationally gentrify in the 1980s.

Buffalo had the highest percentage of central city tracts that slowly gentrified at 22.5%. Providence was only other city in which at least 20% of the central city tracts occupationally gentrified at 20.5%. In addition, there were 13 cities in which at least 10% of the central city experienced slow occupational gentrification. In total, 90 of the 100 cities had at least one central city tract experience slow occupational gentrification in the 1980s.

Jersey City had the highest percentage of central city tracts that experienced rapid occupational gentrification in the 1980s at 37.3%. It was also the only city in which at least 30% of the central city experienced rapid occupational gentrification. There were six cities (Bridgeport, Minneapolis, Gary, Washington, Arlington, and Long Beach) in which at least 20% of the central city tracts rapidly gentrified by occupation. Minneapolis and Washington were among the cities with the five highest levels of rapid occupational gentrification in both the 1970s and the 1980s. Minneapolis was 1st in the 1970s and 3rd in the 1980s, while Washington was 3rd in the 1970s and 5th in the 1980s. In total, 95 of the 100 cities had at least one central city tract experience rapid occupational gentrification during the 1980s.

Turning now to Panel B, which measures the impact of gentrification as the percentage of gentrifiable tracts that experience occupational gentrification, Jersey City and Youngstown had the highest percentage of gentrifiable tracts experiencing occupational gentrification at 66.7%. There were three other cities (Gary, Arlington, and Buffalo) in which at least 60% of the gentrifiable tracts occupationally gentrified and another eight cities in which at least 50% of their gentrifiable tracts occupationally gentrified. Among the cities with the five highest percentages of gentrifiable tracts that occupationally gentrified, only Arlington was also among the top five in the 1970s. Arlington was 3rd in the 1970s and 4th in the 1980s. Buffalo had the highest percentage of gentrifiable tracts that slowly gentrified at 33.3%. The only other city with a percentage above 30% was Providence at 30.8%. There were also 13 cities in which at least 20% of the gentrifiable tracts slowly gentrified by occupation. None of the five cities with the highest percentage of gentrifiable tracts that slowly gentrified in the 1980s were among the top five in the 1970s. Arlington, at 59.1%, had the highest percentage of gentrifiable tracts that rapidly gentrified by occupation in the 1980s. There were three other cities (Jersey City, Youngstown, and Gary) in which at least 50% of the gentrifiable tracts experienced rapid occupational gentrification. In addition, there were two cities (Long

Beach and Seattle) in which at least 40% of the gentrifiable tracts rapidly gentrified by occupation. None of the five cities with the highest percentage of gentrifiable tracts that rapidly gentrified by occupation in the 1980s were among the cities with the highest percentages in the 1970s.

In the 1990s, Newark had the highest percentage of central city tracts that occupationally gentrified, at 40.2%, and was the only city that had at least 40% of its central city tracts occupationally gentrify. There were also three cities (Portland, Cleveland, and Richmond) in which at least 30% of the central city tracts experienced occupational gentrification. None of the cities with the five highest levels of overall occupational gentrification in the 1990s were in the top five in either the 1970s or 1980s. During the 1990s, all 100 cities had at least one central city tract experience occupational gentrification.

Bridgeport had the highest level of slow occupational gentrification with 15.8% of central city tracts slowly gentrifying by occupation. There were three other cities (Miami, Madison, and Milwaukee) in which at least 10% of central city tracts slowly gentrified by occupation. St. Louis was the only one of the five cities with the highest levels of slow gentrification that had been among the top five in previous decades. St. Louis was 3rd in the 1980s and was 5th in the 1990s. During the 1990s, 87 of the 100 cities had at least one central city tract slow gentrify by occupation.

Newark, with 32.2% of central city tracts rapidly gentrifying by occupation, had the highest level of rapid gentrification and was the only city in which at least 30% of central city tracts rapidly gentrified by occupation. There were 14 cities in which at least 20% of central city tracts experience rapid occupational gentrification. Among the cities with the five highest levels of rapid occupational gentrification, only Salt Lake City was among the top five in the previous decades. Salt Lake City was 5th in the 1970s and 3rd in the 1990s. Overall, 97 of the 100 cities had at least one central city tract that experienced rapid occupational gentrification.

Turning to the results in Panel B for the 1990s shows that San Francisco, at 82.1%, had the highest percentage of gentrifiable tracts that occupationally gentrified and was the only city with a percentage above 80%, while Seattle, at 72.6%, was the only other city with a percentage above 70%. There were also six cities (Portland, El Paso, Tulsa, Madison, Pittsburgh, and Lincoln) in which the percentage was at least 60%. Among the five cities with the highest percentages of gentrifiable tracts that occupationally gentrified in the 1990s, only Seattle and Portland were among the cities with the highest percentages in the earlier decades. Seattle was 1st in the 1970s and 2nd in the 1990s, while Portland was 2nd in the 1970s and 3rd in the 1990s.

Madison, at 33.3%, had the highest percentage of gentrifiable tracts that slowly gentrified by occupation in the 1990s and was the only city with at least 30% of the gentrifiable tracts slowly gentrifying. There were four cities (Greensboro, Milwaukee, Lincoln, and Springfield) in which at least 20% of the gentrifiable tracts slowly gentrified by occupation. None of the five cities with the highest percentage of gentrifiable tracts slowly gentrifying by occupation were among the five highest percentages in the earlier decades. San Francisco, at 64.3%, had the highest percentage of gentrifiable tracts that rapidly gentrified by occupation in the 1990s and was the only city in which at least 60% of the gentrifiable tracts rapidly gentrified by occupation. There were six cities (El Paso, Seattle, Tulsa, Portland, Pittsburgh, and Salt Lake City) in which at least 50% of the gentrifiable tracts rapidly gentrified.[2] Among the cities with the five highest percentages of gentrifiable tracts that rapidly gentrified by occupation in the 1990s, only Seattle was among the cities with the highest percentages in the previous decades. Seattle was 1st in the 1970s and 3rd in the 1990s.

The previous tables indicated that there was a large increase in the amount of occupational gentrification during the 2000s and the city-level results reflect this large increase. In the previous decades, the highest percentage of central city tracts experiencing occupational gentrification was Minneapolis in the 1970s at 49.1%. During the 2000s, both Bridgeport at 63.2% and Gary at 58.1% exceeded the previous high by a large amount. In addition, there were five cities (Washington, Atlanta, Tacoma, Newark, and Hartford) in which at least 40% of central city tracts experienced occupational gentrification and another 20 cities with a percentage of at least 30%. Among the cities with the five highest levels of occupational gentrification, only Bridgeport and Washington were among the top five in any of the previous decades. Bridgeport was 3rd in the 1980s and 1st in the 2000s. Washington was 4th in the 1970s and 3rd in the 2000s. All 100 cities had at least one central city tract that experienced occupational gentrification.

Tacoma had the highest percentage of central city tracts slowly gentrifying by occupation at 11.6%. There was only one other city, Hartford at 10.3%, with at least 10% of central city tracts slowly gentrifying by occupation. Three of the five cities with the highest percentage of tracts slowly gentrifying had been among the top five in a previous decade. Hartford was 1st in the 1970s and was 2nd in the 2000s. Providence was 2nd in the 1980s and was 3rd in the 2000s. Baltimore was 2nd in the 1970s and was 4th in the 2000s. Overall, 71 of the 100 cities had at least one central city tract that slowly gentrified by occupation.

Bridgeport had the highest percentage of central city tracts that rapidly gentrified at 63.2%. Gary, at 58.1%, was the only other city in which more than 50% of the central city tracts experienced rapid occupational gentrification. In addition, there were three cities (Washington, Atlanta, and Newark) in which at least 40% of the central city tracts rapidly gentrified by occupation and an additional 15 cities in which at least 30% of the central city tracts experienced rapid occupational gentrification. Among the five cities with the highest percentages of central city tracts rapidly gentrifying by occupation, four of them have been among the top five in previous decades. Bridgeport was 2nd in the 1980s and 1st in the 2000s. Gary was 4th in the 1980s and 2nd in the 2000s. Washington was 3rd in the 1970s, 5th in the 1980s, and 3rd in the 2000s. Finally, Newark was 1st in the 1990s and 5th in the 2000s. All 100 cities had at least one central city tract that rapidly gentrified by occupation in the 2000s.

Turning to Panel B, which measures the impact of gentrification as the percentage of gentrifiable tracts that gentrified in each city, reveals that Fort Lauderdale, at 83.3%, had the highest percentage of gentrifiable tracts that occupationally gentrified in the 2000s. Portland, at 80.3%, was the only other city in which at least 80% of the gentrifiable tracts occupationally gentrified in the 2000s. There were three cities (Atlanta, Washington, and Seattle) in which at least 70% of the gentrifiable tracts and eight cities (Tacoma, Columbus (GA), Denver, El Paso, Gary, Bridgeport, Grand Rapids, and Jersey City) in which at least 60% of the gentrifiable tracts occupationally gentrified in the 2000s. Among the cities with the five highest percentages of gentrifiable tracts that occupationally gentrified in the 2000s, only Portland and Seattle were among the cities with the highest percentage in earlier decades. In fact, both cities were among the cities with the five highest percentages in every decade except for the 1980s. Portland was 2nd in the 1970s, 3rd in the 1980s, and 2nd in the 2000s, while Seattle was 1st in the 1970s, 2nd in the 1990s, and 5th in the 2000s.

Tacoma, at 17.9%, had the highest percentage of gentrifiable tracts that slowly gentrified by occupation in the 2000s. There were seven other cities (Seattle, Tampa, Greensboro, Hartford, Providence, Des Moines, and Richmond) in which at least 10% of the gentrifiable tracts experienced slow occupational gentrification. Among the cities with the five highest percentages of gentrifiable tracts that slowly gentrified by occupation in the 2000s, only Greensboro was among the cities with the highest percentages in the earlier decades. Greensboro was 2nd in the 1990s and 5th in the 2000s.

Fort Lauderdale, at 83.3%, had the highest percentage of gentrifiable tracts that rapidly gentrified by occupation in the 2000s and was the only city with a percentage above 80%. Atlanta, at 72.2%, and Portland, at 72.1%, were the only cities in which at least 70% of the gentrifiable tracts rapidly gentrified by occupation in the 2000s. Finally, there were five cities (Washington, El Paso, Gary, Denver, and Bridgeport) in which at least 60% of the gentrifiable tracts rapidly gentrified by occupation. Among the cities with the five highest percentages in the 2000s, only Portland and El Paso were among the highest percentage in earlier decades. Portland was 5th in the 1990s and 3rd in the 2000s, while El Paso was 2nd in the 1990s and 5th in the 2000s.

Tables 6.8 and 6.9 contain an overview of how the occupational gentrification trends compare to the income and educational gentrification trends documented in Chapters 4 and 5. Table 6.8 compares occupational gentrification levels to income gentrification levels while Table 6.9 compares occupational gentrification and educational gentrification levels.

Looking first at Table 6.8, in the 1970s, the amount of occupational gentrification far exceeded the amount of income gentrification. There were more than 11 times as many occupationally gentrifying tracts in the 1970s as there were income-gentrifying tracts. The difference was most pronounced for rapid gentrification where there were more than 40 times as many tracts that rapidly gentrified by occupation as there were tracts that rapidly gentrified by income.

TABLE 6.8 Comparison of Occupational Gentrification Levels to Income Gentrification
Levels

Total Number of Tracts Gentrifying by Occupation

Total Gentrifying	1970s	1980s	1990s	2000s
All Gentrifiable Tracts	2,460	2,572	2,589	3,459
Low-Income Gentrifiable	2,121	1,942	1,843	2,797
Very Low-income Gentrifiable	339	630	746	662
Slowly Gentrifying				
All Gentrifiable Tracts	586	901	654	257
Low-Income Gentrifiable	507	704	479	217
Very Low-Income Gentrifiable	79	197	175	40
Rapidly Gentrifying				
All Gentrifiable Tracts	1,874	1,671	1,935	3,202
Low-Income Gentrifiable	1,614	1,238	1,364	662
Very Low-Income Gentrifiable	260	433	571	622

Total Number of Tracts Gentrifying by Income

Total Gentrifying	1970s	1980s	1990s	2000s
All Gentrifiable Tracts	222	644	1,221	1,691
Low-Income Gentrifiable	208	356	944	1,412
Very Low-Income Gentrifiable	14	288	277	279
Slowly Gentrifying				
All Gentrifiable Tracts	178	357	963	875
Low-Income Gentrifiable	171	305	744	751
Very Low-Income Gentrifiable	7	52	219	124
Rapidly Gentrifying				
All Gentrifiable Tracts	44	287	258	816
Low-Income Gentrifiable	37	8951	200	661
Very Low-Income Gentrifiable	7	236	58	155

Ratio of Tracts that Gentrify by Occupation to Tracts that Gentrify by Income

Total Gentrifying	1970s	1980s	1990s	2000s
All Gentrifiable Tracts	11.08	3.99	2.12	2.05
Low-income Gentrifiable	10.20	5.46	1.95	1.98
Very Low-Income Gentrifiable	24.21	2.19	2.69	2.37
Slowly Gentrifying				
All Gentrifiable Tracts	3.29	2.52	0.68	0.29
Low-Income Gentrifiable	2.96	2.31	0.64	0.29
Very Low-Income Gentrifiable	11.29	3.79	0.80	0.32
Rapidly Gentrifying				
All Gentrifiable Tracts	42.59	5.82	7.50	3.92
Low-Income Gentrifiable	43.62	24.27	6.82	1.00
Very Low-Income Gentrifiable	37.14	1.83	9.84	4.01

TABLE 6.9 Comparison of Occupational Gentrification Levels to Educational Gentrification Levels

Total Number of Tracts Gentrifying by Occupation

Total Gentrifying	1970s	1980s	1990s	2000s
All Gentrifiable Tracts	2,460	2,572	2,589	3,459
Low-Income Gentrifiable	2,121	1,942	1,843	2,797
Very Low-Income Gentrifiable	339	630	746	662
Slowly Gentrifying				
All Gentrifiable Tracts	586	901	654	257
Low-Income Gentrifiable	507	704	479	217
Very Low-Income Gentrifiable	79	197	175	40
Rapidly Gentrifying				
All Gentrifiable Tracts	1,874	1,671	1,935	3,202
Low-Income Gentrifiable	1,614	1,238	1,364	662
Very Low-Income Gentrifiable	260	433	571	622

Total Number of Tracts Gentrifying by Education

Total Gentrifying	1970s	1980s	1990s	2000s
All Gentrifiable Tracts	1,123	1,680	1,782	2,914
Low-Income Gentrifiable	1,015	1,308	1377	2,359
Very Low-Income Gentrifiable	108	372	405	555
Slowly Gentrifying				
All Gentrifiable Tracts	477	629	684	803
Low-Income Gentrifiable	430	512	514	649
Very Low-Income Gentrifiable	47	117	170	154
Rapidly Gentrifying				
All Gentrifiable Tracts	646	1,051	1,098	2,111
Low-Income Gentrifiable	585	796	863	1,710
Very Low-Income Gentrifiable	61	255	235	401

Ratio of Tracts that Gentrify by Education to Tracts that Gentrify by Income

Total Gentrifying	1970s	1980s	1990s	2000s
All Gentrifiable Tracts	2.19	1.53	1.45	1.19
Low-Income Gentrifiable	2.09	1.48	1.34	1.19
Very Low-Income Gentrifiable	3.14	1.69	1.84	1.19
Slowly Gentrifying				
All Gentrifiable Tracts	1.23	1.43	0.96	0.32
Low-Income Gentrifiable	1.18	1.38	0.93	0.33
Very Low-Income Gentrifiable	1.68	1.68	1.03	0.26
Rapidly Gentrifying				
All Gentrifiable Tracts	2.90	1.59	1.76	1.52
Low-Income Gentrifiable	2.76	1.56	1.58	0.39
Very Low-Income Gentrifiable	4.26	1.70	2.43	1.55

The results also reveal that the overall gap was much higher for very low-income gentrifiable tracts than for low-income gentrifiable tracts. In the 1970s, the ratio for very low-income gentrifiable tracts was 24.2, while the ratio for low-income gentrifiable tracts was 10.2. The difference between the two types of gentrifiable tracts was more pronounced for slowly gentrifying tracts than for rapidly gentrifying tracts. The ratio for slowly gentrifying tracts was 11.3 for very low-income gentrifiable tracts and 3.0 for low-income gentrifiable tracts, while the ratio for rapidly gentrifying tracts was lower in very low-income gentrifiable tracts than for low-income gentrifiable tracts.

During the 1980s, there were still more occupationally gentrifying tracts than income-gentrifying tracts but the differences narrowed substantially. There were almost four times as many occupationally gentrifying tracts as income-gentrifying tracts in the 1980s. The stark differences in the number of rapidly gentrifying tracts also narrowed. There were less than six times as many tracts that rapidly gentrified by occupation as rapidly gentrified by income in the 1980s.

Turning to the results for the two types of gentrifiable tracts reveals that the ratios in the 1980s were much different than the ratios in the 1970s. First, every ratio is much lower in the 1980s than in the 1970s. Second, the decreases were much more pronounced in very low-income gentrifiable tracts than in low-income gentrifiable tracts. For the total amount of gentrification, the ratio in very low-income gentrifiable tracts fell from 24.2 in the 1970s to 2.2 in the 1980s, while the ratio for low-income gentrifiable tracts fell from 10.2 to 5.5. For slowly gentrifying tracts, the ratio for very low-income gentrifiable tracts fell from 11.3 in the 1970s to 3.8 in the 1980s. For low-income gentrifiable tracts, the decrease was from 3.0 in the 1970s to 2.3 in the 1980s. Finally, for rapidly gentrifying tracts, the ratio for very low-income gentrifiable tracts fell from 37.1 in the 1970s to 1.8 in the 1980s, while the ratio for low-income gentrifiable tracts fell from 43.6 in the 1970s to 24.3 in the 1980s.

The difference in the amount of gentrification activity continued to decrease in the 1990s. During the 1990s, there were 2.1 times as many tracts that gentrified by occupation than gentrified by income. However, there was an increase in the gap in the number of tracts that rapidly gentrified. In the 1990s, there were 7.5 times as many tracts that rapidly gentrified by occupation as rapidly gentrified by income compared to 5.8 times in the 1980s.

Separating the results into the two types of gentrifiable tracts reveals some differences in the 1990s. For overall gentrification, the ratio for very low-income gentrifiable tracts increased from 2.2 in the 1980s to 2.7 in the 1990s, while the ratio for low-income gentrifiable fell from 5.5 in the 1980s to 2.0 in the 1990s. For slowly gentrifying tracts, the ratio for very low-income gentrifiable tracts fell from 3.8 in the 1980s to 0.8 in the 1990s, while the ratio for low-income gentrifiable tracts fell from 2.3 in the 1980s to 0.6 in the 1990s. Recall that a ratio below one indicates that there were more tracts that slowly gentrified by income in both types of gentrifiable tracts than gentrified by occupation. Finally, for rapidly gentrifying tracts, the ratio for very low-income gentrifiable tracts increased substantially from 1.8 in the 1980s to 9.8 in the 1990s, while the ratio for low-income gentrifiable tracts fell from 24.3 in the 1980s to 6.8 in the 1990s.

Previously it was documented that there was a large increase in the number of tracts that gentrified along all three dimensions in the 2000s compared to the previous decades. This across-the-board increase in gentrification levels led to the gaps between the different types of gentrification remaining at essentially the same levels in the 2000s as in the 1990s. There were 2.05 times as many tracts that gentrified by occupation as gentrified by income in the 2000s compared to 2.12 in the 1990s. However, while the gap in the total amount of gentrification activity stabilized in the 2000s, the gap with respect to the number of rapidly gentrifying tracts continued to narrow. In the 2000s, there were 3.9 times as many tracts that rapidly gentrified by occupation as rapidly gentrified by income compared to 7.5 times as many in the 1990s.

In the 2000s, the ratios for very low-income gentrifiable tracts fell in each case. For the overall levels of gentrification, the ratio fell from 2.7 in the 1990s to 2.4 in the 2000s. For slowly gentrifying tracts, the ratio fell from 0.8 in the 1990s to 0.3 in the 2000s, while for rapidly gentrifying tracts, the ratio fell from 9.8 in the 1990s to 4.0 in the 2000s. For low-income gentrifiable tracts, the overall ratio was essentially unchanged in the 2000s when compared to the 1990s. However, the ratios for both slowly gentrifying and rapidly gentrifying tracts decreased. For slowly gentrifying tracts, the ratio fell from 0.6 in the 1990s to 0.3 in the 2000s and for rapidly gentrifying tracts, the ratio fell from 6.8 in the 1990s to 1.0 in the 2000s.

Table 6.9, which compares occupational and educational gentrification levels, reveals similar trends to the comparison between occupational and income gentrification. The key differences are that the initial gap is much smaller for educational gentrification than it was for income gentrification. In the 1970s, there were 2.2 times as many tracts that gentrified by occupation as gentrified by education. This gap is much smaller than the gap between income and occupation (more than 11 times). The gap was larger for rapidly gentrifying tracts (2.9) than for slowly gentrifying tracts (1.2) but the gaps are, once again, much smaller than they were for income gentrification.

Separating the results from the 1970s into the two types of gentrifiable tracts reveals that the ratios were slightly higher in very low-income gentrifiable tracts than in low-income gentrifiable tracts. For the overall number of gentrifying tracts, the ratio was 3.1 in very low-income gentrifiable tracts and 2.1 in low-income gentrifiable tracts. For slowly gentrifying tracts, the ratio was 1.7 in very low-income gentrifiable tracts and 1.2 in low-income gentrifiable tracts. Finally, for rapidly gentrifying tracts, the ratio was 4.3 in very low-income gentrifiable tracts and 2.8 in low-income gentrifiable tracts.

During the 1980s, the gap between the amount of occupational gentrification and the amount of educational gentrification narrowed. In the 1970s, there were 2.2 times as many occupationally gentrifying tracts as educationally-gentrifying tracts while in the 1980s, there were 1.5 times as many. Interestingly, the entire decrease was due to a large drop in the gap in the number of tracts that rapidly gentrified. In the 1970s, there were 2.9 times as many tracts that rapidly gentrified by occupation, while in the 1980s, there were only 1.6 times as many. The gap regarding slowly gentrifying tracts increased from 1.2 times in the 1970s to 1.4 times in the 1980s.

Once again, the separate results for the two types of gentrifiable tracts show that very low-income gentrifiable tracts had higher ratios than low-income gentrifiable tracts. However, the differences were smaller in the 1980s than they were in the 1970s. For the overall number of gentrifying tracts, the ratio was 1.7 for very low-income gentrifiable tracts compared to 1.5 for low-income gentrifiable tracts. For slowly gentrifying tracts, the ratio was also 1.7 for very low-income gentrifiable tracts and 1.4 for low-income gentrifiable tracts. Finally, for rapidly gentrifying tracts, the ratio was 1.7 for very low-income gentrifiable tracts and 1.6 for low-income gentrifiable tracts.

In the 1990s, the overall gap in the number of occupationally gentrifying tracts and the number of educationally gentrifying tracts narrowed slightly from 1.53 in the 1980s to 1.45 in the 1990s. This was entirely due to a decrease in the ratio for slowly gentrifying tracts which fell from 1.43 in the 1980s to 0.96 in the 1990s. This means that in the 1990s, there were more tracts that slowly gentrified by education than slowly gentrified by occupation. The ratio for rapid gentrification increased from 1.59 in the 1980s to 1.76 in the 1990s.

Looking at the separate results for the two types of gentrifiable tracts reveals that, once again, the ratios are higher for very low-income gentrifiable tracts than for low-income gentrifiable tracts and that the difference widened slightly in the 1990s when compared to the 1980s. For the overall number of gentrifying tracts, the ratio for very low-income gentrifiable tracts increased from 1.7 in the 1980s to 1.8 in the 1990s, while the ratio for low-income gentrifiable tracts decreased from 1.5 in the 1980s to 1.3 in the 1990s. For slowly gentrifying tracts, the ratio for very low-income gentrifiable tracts fell from 1.7 in the 1980s to 1.0 in the 1990s, while the ratio for low-income gentrifiable tracts fell from 1.4 in the 1980s to 0.9 in the 1990s. Finally, for rapidly gentrifying tracts, the ratio for very low-income gentrifiable tracts increased from 1.7 in the 1980s to 2.4 in the 19909s, while the ratio for low-income gentrifiable tracts remained constant at 1.6.

Finally, in the 2000s, every ratio decreased relative to their levels in the 1990s. The ratio for the total number of gentrifying tracts decreased from 1.45 in the 1990s to 1.19 in the 2000s. The ratio for the number of tracts slowly gentrifying fell from 0.96 in the 1990s to 0.32 in the 2000s. This means that in the 2000s, there were more than three times as many tracts that slowly gentrified by education as slowly gentrified by occupation. Finally, the ratio for rapid gentrification fell from 1.76 in the 1990s to 1.52 in the 2000s.

In the 2000s, the ratios for both types of gentrifiable tracts fell in every case when compared to the 1990s. For the overall number of gentrifying tracts, the ratio for very low-income gentrifiable tracts fell from 1.8 in the 1990s to 1.2 in the

2000s, while for low-income gentrifiable tracts, the ratio fell from 1.3 in the 1990s to 1.2 in the 2000s. For slowly gentrifying tracts, the ratio for very low-income gentrifiable tracts fell from 1.0 in the 1990s to 0.3 in the 2000s, while the ratio for low-income gentrifiable tracts decreased from 0.9 in the 1990s to 0.3 in the 2000s. Finally, for rapidly gentrifying tracts, the ratio for very low-income gentrifiable tracts went from 2.4 in the 1990s to 1.6 in the 2000s and the ratio for low-income gentrifiable tracts decreased from 1.6 in the 1990s to 0.4 in the 2000s.

To summarize the results from Tables 6.8 and 6.9, occupational gentrification was far more prevalent than both income and educational gentrification in the 1970s. Over time, the gap between occupational gentrification and income/educational gentrification narrowed. However, occupational gentrification remained more common than both income and educational gentrification in every decade.

Table 6.10 provides information regarding which cities were most impacted by the combination of income, educational, and occupational gentrification from 1970 to 2010. To identify which cities were most impacted by gentrification, a very simple measure was created. In each decade and for each of the three types of gentrification, the cities were ranked from 1 to 100 with respect to the percentage of central city tracts that rapidly gentrified. As in the previous chapter, if a city did not have any tracts gentrify in a decade, then the city was not given a ranking. Then, each city's average ranking across the three types of gentrification was calculated and is used to identify the cities that were most heavily impacted by gentrification in each decade.

During the 1970s, only Washington, DC, at 7.33, had an average ranking below 10. Denver had the second lowest average score at 10.33, while Oakland had the third lowest average score at 14.00. The rest of the top five consists of San Francisco (15.33) and Chicago (20.0). In addition, there were three cities (Anaheim, Santa Ana, and Warren) in the 1970s that had no tracts rapidly gentrify in any of the three dimensions.

During the 1980s, Jersey City, with an average ranking of 3.0, was, by far, the city with the lowest average ranking across the three types of gentrification. Salt Lake City, with an average ranking of 9.0, was the only other city with an average ranking below 10. Denver had the third lowest average ranking at 19.0 and is the only city in the top three in both the 1970s and the 1980s. The rest of the top five consists of Kansas City (20.0) and Long Beach (20.3). Denver, Oakland, and Boston are the only three cities that were among the ten lowest average rankings in both the 1970s and the 1980s. Warren, MI, was the only city in the 1980s to have no tracts rapidly gentrify along any of the three dimensions in the 1980s and has the distinction of being the only city in the sample with no rapidly gentrifying tracts in either the 1970s or 1980s.

Denver, at 6.33, had the lowest average ranking in the 1990s and was the only city with an average ranking below 10. Atlanta, Chicago, and Richmond (VA) tied for the next lowest average ranking at 10.67. Jersey City (12.67) was the final member of the top five. Denver, Jersey City, and Portland are the only cities among

TABLE 6.10 Average Combined Ranking for Income, Educational, and Occupational Gentrification

Average Ranking for Percentage of Central City Tracts that Rapidly Gentrified by Income, the Percentage that Rapidly Gentrified by Education, and the Percentage that Rapidly Gentrified by Occupation

1970s		1980s		1990s		2000s	
City	*Average*	*City*	*Average*	*City*	*Average*	*City*	*Average*
Washington, DC	7.33	Jersey City, NJ	3.00	Denver, CO	6.33	Atlanta, GA	2.00
Denver, CO	10.33	Salt Lake City, UT	9.00	Atlanta, GA	10.67	Washington, DC	3.67
Oakland, CA	14.00	Denver, CO	19.00	Chicago, IL	10.67	Denver, CO	4.33
San Francisco, CA	15.33	Kansas City, MO	20.00	Richmond, VA	10.67	St. Louis, MO	8.67
Chicago, IL	20.00	Long Beach, CA	23.33	Jersey City, NJ	12.67	Miami, FL	9.67
Oklahoma City, OK	20.33	Seattle, WA	23.33	San Francisco, CA	12.67	Jersey City, NJ	13.00
Shreveport, LA	20.67	Portland, OR	23.67	Dayton, OH	13.33	St. Paul, MN	13.00
Honolulu, HI	22.00	Austin, TX	25.33	Portland, OR	15.33	Salt Lake City, UT	14.33
Philadelphia, PA	26.33	Oakland, CA	25.33	Cleveland, OH	17.33	Chicago, IL	19.00
Detroit, MI	27.33	Boston, MA	25.67	Tulsa, OK	17.33	Cleveland, OH	20.00
Cities with No Gentrifying Tracts	3		1		2		0

the ten lowest average rankings in both the 1980s and the 1990s, while Denver is the only city to be among the ten lowest average rankings in the 1970s, 1980s, and 1990s. In the 1990s, there were two cities (Arlington (VA) and Des Moines) that had no tracts rapidly gentrify in any of the three dimensions. At this point, every city in the sample has had at least one tract rapidly gentrify in one dimension at some point.

In the 2000s, there were five cities with an average ranking below 10. Atlanta had the lowest average ranking at 2.33. This was the lowest average ranking for any city in any decade in the study. Washington had the second-lowest average ranking at 4.00, while Denver had the third-lowest average ranking at 4.33. St. Louis, with an average ranking of 8.67, was the fourth city and Miami, at 9.67, was the fifth city with an average ranking below 10. Atlanta and Denver were among the cities with the three lowest average rankings in both the 1990s and the 2000s, while Jersey City, Chicago, and Cleveland were also among the cities with the ten lowest average rankings in the 1990s and 2000s. In addition, Jersey City, Chicago, and Cleveland were among the lowest 10 average rankings in both the 1980s and the 2000s. Denver and Jersey City were the only cities to be among the 10 lowest average rankings in the 1980s, 1990s, and 2000s, while Denver in the only city to be among the 10 lowest average rankings in every decade in the study. Every city in the sample had at least one tract gentrify in at least one dimension in the 2000s.

Finally, Table 6.11 reports the cities that were most impacted by the various types of gentrification over the entire 1970–2010 timeframe. As was mentioned above, to be eligible to be included in this list, a city had to have at least one tract gentrify in each decade included in the study. The results are presented individually for each type of gentrification as well as for the combined impact of all three types of gentrification. The results for income and educational gentrification were already reported in Table 5.11 but are included here for completeness. However, only the occupation-alone and combined ranking are discussed here.

There were two cities with an average ranking for occupational gentrification below 10. Washington had the lowest average ranking of 5.75, while Denver had the second-lowest average ranking of 9.00. The rest of the top five include Salt Lake City (10.25), St. Louis (11.5) and Atlanta, Jersey City, and St. Paul who all had an average ranking of 13.25.

The final column of Table 6.11 can be interpreted as identifying the cities that were most significantly impacted by the combination of income, educational, and occupational gentrification over the entire period included in this study. From the results, it is clear that Denver stands out as the city with the strongest sustained impact from these types of gentrification. Denver's average ranking across all types of gentrification and across all decades is 10.0 and is the only average ranking below 20. Chicago had the second lowest average ranking of 20.75 and is well ahead of Honolulu, which had the third lowest average ranking of 29.75. The final two cities with the lowest average combined rankings are San Francisco (35.33) and New York (35.42).

TABLE 6.11 Average Ranking Across All Decades for Income, Education, Occupation, and Combined

Income		Education		Occupation		Combined	
City	Average Ranking	City	Average Ranking	City	Average Ranking	City	Average Ranking
Denver, CO	12.5	Salt Lake City, UT	3.5	Washington, DC	5.75	Denver, CO	10
Chicago, IL	19.5	Minneapolis, MN	7.25	Denver, CO	9.00	Chicago, IL	20.75
Oklahoma, OK	20.25	Seattle, WA	7.75	Salt Lake City, UT	10.25	Honolulu, HI	29.75
Kansas City, MO	23.5	Boston, MA	8.5	St. Louis, MO	11.50	San Francisco, CA	35.33
Detroit, MI	26	Denver, CO	8.5	Atlanta, GA	13.25	New York, NY	35.42
Houston, TX	31.5	Jersey City, NJ	9.25	Jersey City, NJ	13.25	Oklahoma City, OK	38.67
Norfolk, VA	31.75	Portland, OR	9.75	St. Paul, MN	13.25	Norfolk, VA	39.33
New York, NY	33.75	Richmond, VA	11	Oakland, CA	14.25	Detroit, MI	40.42
Memphis, TN	37.5	Honolulu, HI	13.75	Minneapolis, MN	15.75	Kansas City, MO	40.92
San Francisco, CA	38.5	St. Paul, MN	14.75	Boston, MA	17.00	Houston, TX	47.08
				Cleveland, OH	17.00		

Conclusion

This chapter has analyzed the trends in occupations gentrification in U.S. cities from 1970–2010. In order to occupationally gentrify, a gentrifiable tract (low-income, central city tract) had to have an increase in its share of employed workers in professional and executive occupation that exceeded the increase in the metropolitan area containing the city. A city is considered to have *slowly* gentrified if its increase simply exceeded the metropolitan area's increase. A city is said to have *rapidly* gentrified if its increase was at least 50% greater than the metropolitan area's increase.

The main results from this chapter are:

- Occupational gentrification was far more common than both income and educational gentrification. The difference was at its greatest in the 1970s and narrowed in each subsequent decade.
- While the number of occupationally gentrifying tracts increased each decade, the increases were much smaller than they were for income and educational gentrification.
- Unlike income and educational gentrification, the probability that a gentrifiable tract occupationally gentrified decreased in the 1980s and 1990s. However, similarly to the other types of gentrification, the probability reached its highest level in the 2000s.
- Denver was identified as the city most affected by the combination of income, educational, and occupational gentrification over the entire 1970–2010 period.

Notes

1 Virginia Beach had 100% of its gentrifiable tracts occupationally gentrify in the 1970s. However, since it only had two gentrifiable tracts, it is excluded from the discussion here.
2 Even though Virginia Beach had 50% of its gentrifiable tracts rapidly gentrify by income, it is omitted from this list because it only had four gentrifiable tracts.

Bibliography

Atkinson, Rowland (2000). "Measuring Gentrification and Displacement in Greater London", *Urban Studies*, 37(1): 149–165.

7

THE VARIETIES OF GENTRIFICATION IN U.S. CITIES, 1970–2010

Each of the previous three chapters analyzed a single dimension of gentrification in U.S. cities from 1970 to 2010. Chapter 4 focused on income gentrification, Chapter 5 focused on educational gentrification, and Chapter 6 focused on occupational gentrification. This chapter will consider gentrification from a multidimensional perspective. Specifically, this chapter will measure how common the different combinations of the three types of gentrification were in each of the decades studied in the previous chapters. For example, this chapter will identify how common it was for tracts to "completely gentrify" which means that they gentrify along all three dimensions. Likewise, the chapter will also identify how common it was for tracts to gentrify only along a single dimension or along any of the combinations of two of the three dimensions. The primary goals are to identify which combinations of the three types of gentrification were the most common in each decade and whether the nature of gentrification in U.S. cities changed over the 40 years included in this study. In addition, the second half of the chapter will break down the results by region and by city to determine if there are any notable geographic patterns with respect to how gentrification was experienced across the United States from 1970 to 2010.

Table 7.1 contains the breakdown of the total number of tracts experiencing gentrification in each decade, as well as the breakdown of the number of tracts experiencing the different varieties of gentrification. The results are reported in two ways. Panel A reports the distribution of the different varieties of gentrification when a tract only needs to meet the criteria for slowly gentrifying to be classified as gentrifying. Panel B reports the distribution when the stricter criteria for rapidly gentrifying is used.

It should be noted that the results in Table 7.1 are presented somewhat differently than the results in the previous three chapters. When focusing only on one type of gentrification, it is possible to distinguish between tracts that slowly gentrified

DOI: 10.1201/9781003217459-7

TABLE 7.1 Total Number of Tracts Experiencing Each Type of Gentrification

Panel A: Total Number of Tracts That Gentrify

	1970s		1980s		1990s		2000s	
	Number	%	Number	%	Number	%	Number	%
Total	2,673		3,207		3,587		4,557	
Complete	132	4.94%	328	10.23%	453	12.63%	956	20.98%
Income Only	41	1.53%	151	4.71%	448	12.49%	300	6.58%
Education Only	174	6.51%	427	13.31%	446	12.43%	637	13.98%
Occupation Only	1,485	55.56%	1,311	40.88%	1,182	32.95%	1,142	25.06%
Income and Education	10	0.37%	77	2.40%	120	3.35%	188	4.13%
Income and Occupation	34	1.27%	85	2.65%	196	5.46%	229	5.03%
Education and Occupation	797	29.82%	828	25.82%	742	20.69%	1,105	24.25%

Panel B: Number of Tracts That Rapidly Gentrify

	1970s		1980s		1990s		2000s	
	Number	%	Number	%	Number	%	Number	%
Total	1,967		2,141		2,367		3,866	
Complete	31	1.58%	96	4.48%	126	5.32%	509	13.17%
Income Only	5	0.25%	119	5.56%	82	3.46%	136	3.52%
Education Only	97	4.93%	328	15.32%	345	14.58%	487	12.60%
Occupation Only	1,321	67.16%	952	44.47%	1,165	49.22%	1,514	39.16%
Income and Education	2	0.10%	37	1.73%	17	0.72%	57	1.47%
Income and Occupation	4	0.20%	34	1.59%	32	1.35%	117	3.03%
Education and Occupation	507	25.78%	575	26.86%	600	25.35%	1,046	27.06%

and those that rapidly gentrified. However, when all three types of gentrification are included, there is the possibility that a tract rapidly gentrifies in one dimension and slowly gentrifies in another. Because of this, a clean distinction between slowly gentrifying and rapidly gentrifying is not possible. So, panel A of Table 7.1 includes all tracts that meet the criteria for slowly gentrifying. This means that it includes the tracts counted as "total gentrifying" in the previous chapters. Panel B includes only tracts that meet the criteria for rapidly gentrifying.

As can be seen from the first row of the table, the number of tracts impacted by gentrification increased each decade. This is true for both slow and rapid gentrification. Using the slow definition of gentrification, the number of tracts impacted by gentrification was 20% higher in the 1980s than in the 1970s, 12% higher in the 1990s than in the 1980s, and 27% higher in the 2000s than in the 1990s. Using the stricter threshold that is required for rapid gentrification, there was a 9% increase in the 1980s relative to the 1970s, an 11% increase in the 1990s relative to the 1980s, and a 63% increase in the 2000s relative to the 1990s. Thus, for both definitions of gentrification, the largest increase in the number of tracts experiencing gentrification relative to the previous decades was for the 2000s relative to the 1990s. This result is consistent with the results from previous chapters which found that there was a surge in gentrification activity in the 2000s.

The rest of Table 7.1 provides the number of tracts experiencing the various combinations of gentrification activity in each decade. With three different types of gentrification (income, educational, and occupational), there are seven different combinations of gentrification activity that a tract can experience:

- Complete (gentrifies along all three dimensions)
- Income Only
- Education Only
- Occupation Only
- Income and Education
- Income and Occupation
- Education and Occupation

During the 1970s, using the less restrictive, slow definition of gentrification, the most common form of gentrification, by far, was Occupation Only. Over half of the tracts (56%) that experienced any sort of gentrification activity in the 1970s gentrified only by occupation. The next most common form of gentrification was the combination of Education and Occupation. An additional 30% of the tracts that gentrified during the 1970s did so for both education and occupation. Thus, more than 86% of the tracts that gentrified in the 1970s experienced one of two types of gentrification: Occupation Only or Education-Occupation. The least common form of gentrification was Income-Education with only 0.4% of the tracts that gentrified experiencing that form of gentrification. These results are consistent with what was learned from the previous three chapters. Chapter 4 showed that income gentrification was quite rare in the 1970s, while the number of tracts that occupationally gentrified far exceeded the number that gentrified along either of the other two dimensions.

Switching to the more restrictive rapid definition of gentrification yields qualitatively similar results. However, rapid gentrification in the 1970s is even more concentrated in the categories of Occupation Only and Education-Occupation. Among the tracts that rapidly gentrified in some way, 67% of them gentrified by

Occupation Alone and 26% gentrified by Education and Occupation. Thus, about 93% of the rapid gentrification activity in the 1970s was in these two categories. The least common form of rapid gentrification in the 1970s was Income-Education with only two tracts experiencing this form of gentrification.

The most common forms of gentrification remained the same in the 1980s for both slow and rapid gentrification. However, the combined share of gentrification activity for Occupation Only and Education-Occupation decreased substantially for both types of gentrification. Using the slow definition of gentrification, the percentage of gentrifying tracts that gentrified by Occupation Only decreased from 56% in the 1970s to 41% in the 1980s. Likewise, the percentage of gentrifying tracts that gentrified by both Education and Occupation fell from 30 to 26%. Thus, their combined share of gentrifying tracts decreased from 85 to 67%. The biggest increases in the share of gentrifying tracts were for Education Only, which increased by 6.8 percentage points and Complete Gentrification which increased by 5.3 percentage points.

Switching to the results in panel B, which use the rapid definition of gentrification, the percentage of gentrifying tracts that gentrified by Occupation Only fell from 67% in the 1970s to 44% in the 1980s and the percentage that gentrified by Education and Occupation increased slightly from 24 to 26%. Thus, their combined share fell from 91 to 71%. The largest increase in the share of rapidly gentrifying tracts was for Education Only where the share of rapidly gentrifying tracts increased by 10.4 percentage points from 4.9% in the 1970s to 15.3% in the 1980s.

Turning now to the change in the number of tracts that experienced the various types of slow gentrification from the 1970s to the 1980s reveals that there was only one type of gentrification (Occupation Only) that had fewer tracts experiencing it in the 1980s than in the 1970s. There were 174 fewer tracts that gentrified by Occupation Alone in the 1980s. The largest increases in the number of tracts experiencing a particular type of gentrification were for Education Only (+253) and Complete (+196).

The results for the change in the number of tracts that experienced the various types of rapid gentrification are very similar to those for slow gentrification. Only one type of gentrification (Occupation Only) was experienced by fewer tracts in the 1980s than in the 1970s. The number of rapidly gentrifying tracts that gentrified by occupation alone decreased from 1,321 in the 1970s to 952 in the 1980s. The biggest increase was for Education Only. The number of rapidly gentrifying tracts that gentrified by education alone increased from 97 in the 1970s to 231 in the 1980s.

Occupation Only and Education-Occupation remained the two most common forms of gentrification in the 1990s for both slow and rapid gentrification. For slowly gentrifying tracts, the share of tracts that were either Occupation Only or Education-Occupation continued to decline. During the 1980s, 67% of slowly gentrifying tracts were either Occupation Only or Education and Occupation. During the 1990s, the combined percentage fell to 54%. The percentage of slowly gentrifying tracts that gentrified by Occupation Only decreased by 7.9 percentage points from 40.9 to 33.0% and the percentage of slowly gentrifying tracts that gentrified by both education and occupation decreased from 25.8 to 20.7%.

For rapidly gentrifying tracts, the percentage of gentrifying tracts that were either Occupation Only or Education-Occupation increased from 71.3% in the 1980s to 74.6% in the 1990s. The increase was entirely due to a 4.8-percentage point increase in the percentage of rapidly gentrifying tracts that gentrified by Occupation Only. The increase in the percentage of rapidly gentrifying tracts that gentrified by occupation alone was the largest increase for any of the categories of gentrification.

Regarding the absolute number of tracts experiencing the various types of gentrification, the largest increase in the 1990s relative to the 1980s for slowly gentrifying tracts was for Income Only. There were 297 more tracts that gentrified by Income Only in the 1990s compared to the number that gentrified by income alone in the 1980s. There were two types of gentrification for which there were fewer tracts experiencing that type of gentrification in the 1990s than in the 1980s. The number of tracts gentrifying by occupation alone decreased by 129 in the 1990s relative to the 1980s and the number of tracts gentrifying by both education and occupation decreased by 86.

For rapidly gentrifying tracts, the largest increase in the number of tracts experiencing a particular type of gentrification was for Occupation Only. There were 213 more tracts that rapidly gentrified by occupation only in the 1990s than in the 1980s. The biggest decreases were for Income Only, which saw a decrease of 37 tracts, and Income-Education, which saw a decrease of 20 tracts.

As was the case for every decade so far, the most common types of slowly gentrifying tracts in the 2000s were Occupation Only and Education-Occupation. However, the percentage of slowly gentrifying tracts that were Occupation Only reached its lowest level at 25.1%. This was 7.9 percentage points lower than the share from the 1990s. The percentage of slowly gentrifying tracts that gentrified by both education and occupation increased from 20.7% in the 1990s to 24.3% in the 2000s, an increase of 3.6 percentage points. The combined share for these two types of gentrification fell from 53.6% in the 1990s to 49.3% in the 2000s, the lowest combined share for any decade. The biggest increase in share of slowly gentrifying tracts was for Completely Gentrified tracts. The percentage of slowly gentrifying tracts that gentrified along all three dimensions increased from 12.6% in the 1990s to 21.0% in the 2000s, an increase of 8.4 percentage points. This was also the first time that any gentrification type other than Occupation Alone or Education-Occupation achieved a share above 20%.

For rapidly gentrifying tracts in the 2000s, Occupation Alone and Education-Occupation continue to be the dominant forms of gentrification. Occupation Alone was once again the most common form of rapid gentrification with 39.2% of rapidly gentrifying tracts in the 2000s gentrifying only by occupation. However, the share of rapidly gentrifying tracts that were Occupation Only decreased by 10.1 percentage points from 49.2% in the 1990s to 39.2% in the 2000s. Education and

Occupation also remained the second most common form of rapid gentrification with 27.1% of rapidly gentrifying tracts in the 2000s gentrifying by both education and occupation. This was an increase of 1.7 percentage points from the 1990s when 25.4% of rapidly gentrifying tracts gentrified by both education and occupation. The combined share for the two most common forms of gentrification was 66.2% which was the lowest combined share for any decade and was 8.4 percentage points lower than the combined share for the 1990s. As was the case for slowly gentrifying tracts in the 2000s, the largest increase in the share of rapidly gentrifying tracts was for tracts that completely gentrified. The percentage of rapidly gentrifying tracts that gentrified along all three dimensions increased by 7.8 percentage points from 5.3% in the 1990s to 13.2% in the 2000s.

The results in Table 7.1 show that the most common way for a tract to gentrify in the four decades covered by this study was to gentrify only by occupation while the second most common way for tracts to gentrify was by both occupation and education. However, these two forms became less dominant as time passed. In the 1970s, 85% of gentrifying tracts using the slow definition of gentrification and almost 93% of gentrifying tracts using the rapid definition gentrified either by occupation only or by both occupation and education. In the 2000s, the combined share for these two types of gentrification fell to 49% using the slow definition and 66% using the rapid gentrification. While Occupation Only and Occupation-Education gentrification became much less common between the 1970s and the 2000s, complete gentrification became much more common. In the 1970s, only 4.9% of slowly gentrifying and 1.6% of rapidly gentrifying tracts completely gentrified. In the 2000s, the percentage had risen to 21.0% of slowly gentrifying and 13.2% of rapidly gentrifying tracts. The second largest increase was for tracts that gentrified by Education Only. Among slowly gentrifying tracts, the percentage that gentrified by education alone increased from 6.5% in the 1970s to 14.0% in the 2000s, while the percentage of rapidly gentrifying tracts that gentrified by education only increased from 4.9% in the 1970s to 12.6% in the 2000s.

Finally, the results in Table 7.1 reveal that the types of gentrification experienced in U.S. cities became more varied between 1970 and 2010. In the 1970s, among both slowly and rapidly gentrifying tracts, there were only two types of gentrification (Occupation Alone and Occupation Education) experienced by at least 10% of gentrifying tracts. In the 2000s, for both slow and rapid gentrification, there were four types of gentrification experienced by at least 10% of gentrifying tracts: Occupation Only, Education-Occupation, Complete, and Education Only.

Table 7.2 provides the breakdown of both slow and rapid gentrification by Census region for each decade. Panel A of Table 7.2 contains the results for slow gentrification, while panel B which contains the results for rapid gentrification will be discussed.

Starting with the results in panel A, in the 1970s, for every Census region, the most common form of slow gentrification was Occupation Only and the next most common form was Education-Occupation. The percentage of slowly gentrifying

TABLE 7.2 Type of Gentrification by Census Region

Panel A: Tracts Identified as Gentrifying Using the Criteria for Slowly Gentrifying

	1970s		1980s		1990s		2000s	
Northeast	Number	%	Number	%	Number	%	Number	%
Total	593		992		942		1,251	
Complete	27	4.55%	83	8.37%	97	10.30%	264	21.10%
Income Only	6	1.01%	18	1.81%	75	7.96%	54	4.32%
Education Only	33	5.56%	116	11.69%	141	14.97%	178	14.23%
Occupation Only	362	61.05%	472	47.58%	344	36.52%	306	24.46%
Income and Education	2	0.34%	13	1.31%	38	4.03%	56	4.48%
Income and Occupation	1	0.17%	16	1.61%	46	4.88%	50	4.00%
Education and Occupation	162	27.32%	274	27.62%	201	21.34%	343	27.42%
Midwest								
Total	783		905		999		1,171	
Complete	31	3.96%	81	8.95%	109	10.91%	206	17.59%
Income Only	16	2.04%	35	3.87%	154	15.42%	110	9.39%
Education Only	45	5.75%	119	13.15%	102	10.21%	170	14.52%
Occupation Only	435	55.56%	422	46.63%	351	35.14%	312	26.64%
Income and Education	2	0.26%	29	3.20%	27	2.70%	56	4.78%
Income and Occupation	11	1.40%	26	2.87%	69	6.91%	66	5.64%
Education and Occupation	243	31.03%	193	21.33%	187	18.72%	251	21.43%
South								
Total	684		732		934		1,163	
Complete	61	8.92%	102	13.93%	137	14.67%	264	22.70%
Income Only	7	1.02%	62	8.47%	126	13.49%	91	7.82%
Education Only	45	6.58%	96	13.11%	104	11.13%	163	14.02%
Occupation Only	371	54.24%	258	35.25%	324	34.69%	298	25.62%
Income and Education	3	0.44%	22	3.01%	24	2.57%	39	3.35%
Income and Occupation	13	1.90%	33	4.51%	49	5.25%	51	4.39%
Education and Occupation	184	26.90%	159	21.72%	170	18.20%	257	22.10%

(*Continued*)

TABLE 7.2 (Continued)

Panel A: Tracts Identified as Gentrifying Using the Criteria for Slowly Gentrifying

Northeast	1970s Number	%	1980s Number	%	1990s Number	%	2000s Number	%
West								
Total	615		578		712		972	
Complete	15	2.44%	62	10.73%	110	15.45%	222	22.84%
Income Only	12	1.95%	36	6.23%	93	13.06%	45	4.63%
Education Only	51	8.29%	96	16.61%	99	13.90%	126	12.96%
Occupation Only	317	51.54%	159	27.51%	163	22.89%	226	23.25%
Income and Education	3	0.49%	13	2.25%	31	4.35%	37	3.81%
Income and Occupation	9	1.46%	10	1.73%	32	4.49%	62	6.38%
Education and Occupation	208	33.82%	202	34.95%	184	25.84%	254	26.13%

Panel B: Tracts Identified as Gentrifying Using the Criteria for Rapidly Gentrifying

Northeast	1970s Number	%	1980s Number	%	1990s Number	%	2000s Number	%
Total	1,967		2,141		2,367		3,866	
Complete	31	1.58%	96	4.48%	126	5.32%	509	13.17%
Income Only	5	0.25%	119	5.56%	82	3.46%	136	3.52%
Education Only	97	4.93%	328	15.32%	345	14.58%	487	12.60%
Occupation Only	1,321	67.16%	952	44.47%	1,165	49.22%	1,514	39.16%
Income and Education	2	0.10%	37	1.73%	17	0.72%	57	1.47%
Income and Occupation	4	0.20%	34	1.59%	32	1.35%	117	3.03%
Education and Occupation	507	25.78%	575	26.86%	600	25.35%	1,046	27.06%
Midwest								
Total	408		651		645		1,055	
Complete	4	0.98%	14	2.15%	27	4.19%	116	11.00%
Income Only	0	0.00%	1	0.15%	16	2.48%	20	1.90%
Education Only	20	4.90%	80	12.29%	112	17.36%	128	12.13%
Occupation Only	276	67.65%	371	56.99%	332	51.47%	429	40.66%
Income and Education	0	0.00%	4	0.61%	7	1.09%	10	0.95%
Income and Occupation	1	0.25%	2	0.31%	6	0.93%	28	2.65%
Education and Occupation	107	26.23%	179	27.50%	145	22.48%	324	30.71%

(*Continued*)

TABLE 7.2 (Continued)

Panel B: Tracts Identified as Gentrifying Using the Criteria for Rapidly Gentrifying

	1970s		1980s		1990s		2000s	
Northeast	Number	%	Number	%	Number	%	Number	%
South								
Total	566		581		611		984	
Complete	3	0.53%	23	3.96%	38	6.22%	133	13.52%
Income Only	2	0.35%	39	6.71%	18	2.95%	62	6.30%
Education Only	35	6.18%	87	14.97%	61	9.98%	142	14.43%
Occupation Only	366	64.66%	275	47.33%	354	57.94%	385	39.13%
Income and Education	0	0.00%	16	2.75%	3	0.49%	25	2.54%
Income and Occupation	0	0.00%	20	3.44%	10	1.64%	44	4.47%
Education and Occupation	160	28.27%	121	20.83%	127	20.79%	193	19.61%
West								
Total	507		488		631		991	
Complete	21	4.14%	35	7.17%	39	6.18%	158	15.94%
Income Only	3	0.59%	59	12.09%	23	3.65%	44	4.44%
Education Only	25	4.93%	76	15.57%	79	12.52%	114	11.50%
Occupation Only	328	64.69%	167	34.22%	321	50.87%	381	38.45%
Income and Education	0	0.00%	11	2.25%	4	0.63%	17	1.72%
Income and Occupation	3	0.59%	7	1.43%	10	1.58%	27	2.72%
Education and Occupation	127	25.05%	133	27.25%	155	24.56%	250	25.23%

tracts that gentrified by Occupation Alone ranged from 51.5% in the West to 61.1% in the Northeast, while the percentage of tracts that gentrified by both education and occupation ranged from 26.9% in the South to 33.8% in the West. The combined percentage for these two types of gentrification exceeded 80% in every region. Finally, the least common form of gentrification in every region was Income Only with the percentage ranging from 1.0% in both the Northeast and South to 2.0% in the Midwest. As has been previously established, income gentrification was very rare in the 1970s.

During the 1980s, Occupation Only and Education-Occupation remained the two most common forms of slow gentrification in all four regions. However, for cities in the West, Education-Occupation surpassed Occupation Only as the most common form of gentrification. For cities in the West, 35.0% of the slowly gentrifying tracts gentrified in both education and occupation while 27.5% of them gentrified by occupation alone. For the other three regions, Occupation Only remained the most common form of gentrification with the percentage of slowly gentrifying tracts that gentrified by only occupation ranging 35.3% in the South to 46.6% in the Midwest. Education-Occupation was once again the second most common form of slow gentrification for these regions. The percentage of slowly gentrifying tracts that gentrified by both education and occupation ranged from 21.3% in the Midwest to 27.6% in the Northeast.

The combined share of Occupation Only and Education-Occupation among slowly gentrifying tracts fell in all four regions. The combined share ranged from 57.0% in the South to 75.2% in the Northeast. The largest decline in the percentage of slowly gentrifying tracts that were Occupation Only or Education-Occupation was in the South where the percentage fell by 24.2 percentage points, while the smallest decline was in the Northeast where the share fell by 13.2 percentage points.

The least common forms of slowly gentrifying tracts were essentially the same for all four regions in the 1980s. In the Northeast, there were three types of gentrification that had shares below 2%. The least common form was Income-Education with only 1.3% of slowly gentrifying tracts in the Northeast. The other two forms below 2% were Income-Occupation (1.6%) and Income Only (1.8%). In the Midwest, the same three types of gentrification had shares below 4%. The least common form was Income-Occupation (2.9%), while Income-Education (3.2%) and Income Only (3.9%) were also below 4%. For cities in the South, the same three types of gentrification are the least common. However, a key difference is that the percentage of slowly gentrifying tracts that gentrify by income alone (8.5%) is much higher than for the Northeast and Midwest. The least common form of slow gentrification in the South is Income-Education with 3.0% of slowly gentrifying tracts, while Income-Occupation had a 4.5% share. Finally, the West is similar to the South in that, while the three most common forms of slow gentrification are the same, the share of slowly gentrifying tracts that are Income Only (6.2%) is higher than in the Northeast and the Midwest. Income-Occupation is the least common form of slow gentrification in the West with a share of 1.7% of slowly gentrification tracts. Income-Education had a share of 2.3% of slowly gentrifying tracts in the West.

Clearly, the common thread for the least common forms of gentrification in the 1980s is that they all include income gentrification. As was the case in the 1970s, income gentrification in the 1980s was rare relative to the other two forms of gentrification.

Looking at the changes from the 1970s to the 1980s, for all four regions, the largest decline in the share of slowly gentrifying tracts was for Occupation Only. The decrease ranged from 8.9 percentage points in the Midwest to 24.0 percentage points in the West. For three of the four regions, the largest increase in the share of slowly gentrifying tracts from the 1970s to the 1980s was for Education Only. Among these three regions, the increase in the share of slowly gentrifying tracts that gentrified by education only ranged from 6.1 percentage points in the

Northeast to 8.3 percentage points in the West. For cities in the South, the largest increase (7.5 percentage points) was for Income Only, while Education Only had the second largest increase of 6.5 percentage points.

Turning to the 1990s, Occupation Only was once again the most common form of slow gentrification in three of the four regions. In each of these regions, Education-Occupation was the next most common form of slow gentrification. As was the case in the 1980s, Education-Occupation was the most common form of slow gentrification in the West. Occupation Only was the next most common form in the West. The share of slowly gentrifying tracts that gentrified by occupation alone ranged from 22.9% in the West to 36.5% in the Northeast. The share of slowly gentrifying tracts that gentrified by both education and occupation ranged from 18.2% in the South to 25.8% in the West.

For all four regions, the least common form of slow gentrification in the 1990s was Income-Education, while Income-Occupation was the next least common form. The percentage of slowly gentrifying tracts that gentrified by both income and education ranged from 2.6% in the South to 4.4% in the West, while the percentage of slowly gentrifying tracts that gentrified by both income and occupation ranged from 4.5% in the West to 6.9% in the Midwest.

Looking at the changes from the 1980s to the 1990s reveals that the type of gentrification that had the largest increase in the share of slowly gentrifying tracts was Income Only in all four regions. The increase ranged from 5.0 percentage points in the South to 11.6 percentage points in the Midwest. The regions differed with respect to the type of gentrification that experienced the largest decrease in its share of slowly gentrifying tracts. In the Northeast and Midwest, the largest decrease was for Occupation Only with decreases of 11.1 percentage points in the Northeast and 11.5 percentage points in the Midwest. In both the South and the West, the largest decrease was for Education-Occupation. In the South, the percentage of slowly gentrifying tracts that gentrified in both education and occupation fell by 3.5 percentage points relative to the 1980s, while, in the West, the share decreased by 9.1 percentage points.

Turning to the 2000s, the one thing that stands out is the extent to which the distribution of gentrification types is more evenly distributed than in any of the other decades. Occupation Only remained the most common type of slow gentrification in the Midwest (26.6%) and the South (25.6%), while Education-Occupation was the most common form in the Northeast (27.4%) and the West (26.1%). Interestingly, three of the four regions (Northeast, South, and West) had at least three types of gentrification with shares above 20% in the 2000s, while the Midwest had two types of gentrification with a share above 20% and a third with a share of 17.6%. In the previous three decades, there were no cases in which a region had three types of gentrification with a share above 20%.

The most common form of slow gentrification varies by region. For the Northeast and the West, the most common form of slow gentrification in the 2000s was Education-Occupation with a share of 27.4% in the Northeast and 26.1% in the

West. For both the Midwest and the South, the most common form of gentrification was Occupation Only with a share of 26.6% in the Midwest and 25.6% in the South. For three of the four regions (Midwest, South, and West), the least common form of gentrification was Income-Education with a share that ranged from 3.4% in the South to 4.8% in the Midwest. For the Northeast, the least common form of gentrification was Income-Occupation with a share of 4.0%.

Turning now to the changes in the share of slow gentrification from the 1990s to the 2000s reveals that the biggest increase in the share of slowly gentrifying tracts was for the percentage of tracts that completely gentrified. The increase in the percentage of slowly gentrifying tracts that gentrified along all three dimensions ranges from 6.7 percentage points in the Midwest to 10.8 percentage points in the Northeast. For three of the four regions (Northeast, Midwest, and South), the large decline in the share of slowly gentrifying tracts was for Occupation Only with a decline that ranged from 8.5 percentage points in the Midwest to 12.1 percentage points in the Northeast. For the West, the largest decline in share was for Income Only with a decrease of 8.4 percentage points.

The analysis will now turn to the distribution of rapid gentrification among the various types of gentrification. These results are found in panel B of Table 7.2. In the 1970s, the rapid gentrification activity was dominated by Occupation Only and Education-Occupation. For all four regions, the most common form of rapid gentrification in the 1970s was Occupation Only with a share that ranged from 64.7% in the South and the West to 67.7% in the Midwest. Education-Occupation was the second most common form of gentrification in all four regions as well. The percentage of rapidly gentrifying tracts in the 1970s that gentrified by both education and occupation ranged from 25.1% in the West to 28.3% in the South. The combined share of Occupation Only and Education-Occupation ranged from 89.7% in the West to 93.9% in the Midwest.

Three of the four regions in the 1970s had at least one type of gentrification that was experienced by no tracts. The exception was the Northeast where every type of gentrification was experienced by at least one tract. In the Midwest, there were no tracts that experienced Income Only or Income-Education; in the South, there were no tracts that experienced Income-Education or Income-Occupation, and in the West, there were no tracts that experienced Income-Occupation.

In the 1980s, Occupation Only and Education-Occupation remained the two most common forms of rapid gentrification. However, rapid gentrification in the 1980s was much less dominated by these two forms than in the 1970s. The percentage of rapidly gentrifying tracts that gentrified only by occupation ranged from 34.32% in the West to 57.0% in the Northeast. Once again, Occupation-Only was the most common form of rapid gentrification, while Education-Occupation was the second most common form of rapid gentrification in all four regions. The percentage of rapidly gentrifying tracts that gentrified by both education and occupation ranged from 20.8% in the South to 27.5% in the Midwest. The combined shares of Occupation Only and Education-Occupation ranged from 61.5% in the West to 84.5% in the Midwest.

The least common form of rapid gentrification in the 1980s varied from region to region. In the Northeast, the least common form was Income-Education with a share of 1.7%. For the Midwest, the least common form was Income Only with a share of 0.2%. Income-Education was the least common form in the South with a share of 2.8%, while Income-Occupation was the least common form in the West with a share of 1.4% of rapidly gentrifying tracts.

Comparing the changes in the shares of rapidly gentrifying tracts in the 1980s to those from the 1970s reveals a very large drop in the percentage of gentrifying tracts that gentrified only by occupation. The decrease in the share of Occupation Only ranged from 10.7 percentage points in the Midwest to 30.5 percentage points in the West. For three of the four regions (Northeast, Midwest, and South), the largest increase in share relative to the 1970s was for Education Only. In the West, the largest increase was for Income Only with an increase of 11.5 percentage points. Interestingly, the West also experiences a large increase in the percentage of gentrifying tracts that gentrified by Education Only. During the 1980s, gentrification in the West shifted away from Occupation Only to Income-Only and Education-Only gentrification.

Occupation Only and Education-Occupation remained the two most common forms of rapid gentrification in the 1990s. The share for Occupation Only ranged from 49.2% in the Northeast to 57.9% in the South. Education-Occupation remained the second most common form of gentrification in all four regions with a share ranging from 20.8% in the South to 25.4% in the Northeast. The combined share for Occupation Only and Education-Occupation ranged from 74.0% in the Midwest to 78.7% in the South. Interestingly, the combined share rose in three of the regions relative to the 1980s. This is very different from the results for slow gentrification where the combined share for these two types of gentrification declined in every decade when compared to the previous decade.

Income-Education was the least common form of rapid gentrification in three of the four regions. The exception was the Midwest where Income-Occupation was the least common form. The share for Income-Education ranged from 0.5% in the South to 1.1% in the Midwest, while the share for Income-Occupation ranged from 0.9% in the Midwest to 1.6% in the South. Both types of gentrification were quite rare in the 1990s.

Comparing the change in rapid gentrification shares from the 1980s to the 1990s reveals a lot of variation across regions. For the Northeast, the largest decrease in rapid gentrification share was 2.1 percentage points for Income Only, while the largest increase was 4.8 percentage points for Occupation Only. In the Midwest, the largest decrease was 5.5 percentage points for Occupation Only, while the largest increase was 5.1 percentage points for Education Only. In the South, the largest decrease was 5.0 percentage points for Education Only, while the largest increase was 10.6 percentage points for Occupation Only. Finally, for the West, the largest decrease in share was 8.5 percentage points for Income Only, while the largest increase was 16.7 percentage points for Occupation Only. What is most striking about these results is the surge in Occupation-Only gentrification in the South and the West. This is very different than the results for slow gentrification where the share for Occupation Only in every region was at its maximum in the 1970s and decreased in each subsequent decade.

Turning now to the rapid gentrification results in the 2000s reveals that, once again, Occupation Only and Education-Occupation are the two most common forms of gentrification in every region. Occupation Only is, once again, the most common form in all four regions. The share of rapidly gentrifying tracts that gentrified only by occupation was remarkably similar in all four regions. The share ranged from 38.5% in the West to 40.7% in the Midwest. The share for Education-Occupation ranged from 19.6% in the South to 30.7% in the Midwest. The combined share for Occupation Only and Education-Occupation ranged from 58.7% in the South to 71.4% in the Midwest. The least common form of rapid gentrification in the 2000s was Income-Education in all four regions. The share ranged from 1.0% in the Midwest to 2.5% in the South.

Comparing the gentrification shares from the 1990s to those from the 2000s reveals that the biggest increase in share was for Complete gentrification in three of the four regions. The exception was the Midwest where the 8.2-percentage point increase for Education-Occupation exceeded the 6.8-percentage point increase for Complete gentrification. Among the other three regions, the increase in the percentage of tracts that gentrified by income, education, and occupation ranged from 7.3 percentage points in the South to 9.8 percentage points in the West. The share for Complete gentrification more than doubled when compared to the 1990s in all four regions. As was the case for slow gentrification, one of the striking characteristics of rapid gentrification in the 2000s was a very large increase in the number of tracts that gentrified along all three dimensions.

The largest decrease in the share of rapidly gentrifying tracts in the 2000s relative to the 1990s was Occupation Only for all four regions. The decrease ranged from 10.1 percentage points in the Northeast to 18.8 percentage points in the South.

The results in Table 7.2 are quite similar to the overall results presented in Table 7.1 but do reveal some differences between the regions regarding the types of gentrification that are most and least common. Occupation Only was the most common form of gentrification in three of the four regions in the 1970s, 1980s, and 1990s. The lone exception is that Education-Occupation is the most common form in the West in the 1980s and 1990s. In the 2000s, Education-Occupation remained the most common form in the West and became the most common form in the Northeast. Occupation-Only remained the most common form in the Midwest and the South. The biggest increase in every region from the 1970s to the 2000s was for Complete gentrification.

The final section of this chapter will analyze which cities were impacted the most by the various types of gentrification. As was the case in the previous chapter, the results are presented only for tracts that gentrified rapidly and the impact of gentrification on a city is measured as the percentage of central city tracts that experienced a particular type of gentrification.

In the 1970s, Houston experienced the highest level of Complete gentrification. In Houston, 2.5% of central city tracts gentrified by income, education, and occupation. Fort Lauderdale (2.1%) and Shreveport (1.8%) had the next highest levels of Complete gentrification.

Among tracts that only gentrified along one of the three dimensions, Virginia Beach (1.0%) had the highest level of Income-Only gentrification, Arlington (5.1%) had the highest level of Education-Only gentrification, and Minneapolis (25.9%) had the highest level of Occupation-Only gentrification.

Finally, for the tracts that gentrified along two of the three dimensions in the 1970s, Denver was the only city to experience Income-Education gentrification and had 1.5% of its central city tracts gentrify by both income and education. Shreveport had the highest level of Income-Occupation gentrification at 3.5%, while Minneapolis had the highest level of Education-Occupation gentrification with 18.1% of its central city tracts gentrifying along those two dimensions.

In the 1980s, Jersey City had the highest level of Complete gentrification with 6.0% of its central city tracts gentrifying by income, education, and occupation. Fort Worth (4.0%) and Norfolk (3.8%) had the next highest levels of Complete gentrification.

Among the tracts that gentrified only along one dimension in the 1980s, Madison had the highest levels of both Income-Only and Education-Only gentrification. In Madison, 9.8% of the central city tracts gentrified by Income Only and 13.1% of the central city tracts gentrified by Education Only. Bridgeport had the highest level of Occupation-Only gentrification with 23.7% of its central city tracts gentrifying only by occupation in the 1980s.

Among the tracts that gentrified by two of the three dimensions, Wichita (3.9%) had the highest level of Income-Education gentrification, Lincoln (6.1%) had the highest level of Income-Occupation gentrification, and Jersey City (25.4%) had the highest level of Education-Occupation gentrification.

Turning to the 1990s. Chicago, at 3.7%, had the highest level of Complete gentrification with Denver, at 3.5%, a close second. San Francisco was third with 3.1% of its central city tracts gentrifying by income, education, and occupation.

Among tracts that gentrified only along a single dimension, Shreveport, at 7.0%, had the highest percentage of its central city gentrify by Income Only, Providence (10.3%) had the highest percentage of central city tracts gentrify by Education Only, and Newark (31.0%) had the highest percentage of its central city gentrify by Occupation alone.

For tracts that gentrified along two of the three dimensions in the 1990s, Shreveport had the highest percentage of its central city tracts gentrify by income and education with 1.8% of its central city tracts gentrifying along these two dimensions. Anaheim, at 1.7%, was a close second. Youngstown had the highest level of Income-Occupation gentrification with 3.2% of its central city tracts gentrifying alone these two dimensions. Salt Lake City, at 18.9%, had the highest percentage of its central city tracts gentrify by education and occupation.

The previous three chapters revealed that gentrification levels were much higher in the 2000s than in any of the previous decades. This is also seen in the much higher percentage of tracts that experience Complete gentrification in the 2000s. Atlanta had the highest percentage of its central city tracts completely gentrify with 20.9% of its central city tracts gentrifying by income, education, and occupation in the 2000s. Denver, at 14.2%, had the next highest level, while Akron, at 9.8%, had the third-highest level. In the previous decades, the highest percentage was 6.0% for Jersey City in the 1980s.

Among the tracts that only gentrified along one dimension, Evansville had the highest percentage of its central city gentrify by income alone at 6.8%. Salt Lake City, at 13.2%, had the highest percentage of its central city gentrify by Education alone, while Gary had the highest percentage of its central city gentrify by Occupation alone at 51.6%. The level of Occupation-Only gentrification in Gary in the 2000s far exceeded the levels experienced by any cities in the previous three decades. The highest levels from earlier decades was 31.0% for Newark in the 1990s.

Finally, among the tracts that gentrified along two dimensions, Akron and Madison had the highest percentage of their central cities gentrify by income and education at 3.3%. Akron, at 4.9%, had the highest percentage of its central city gentrify by Income and Occupation, while Washington had the highest percentage of its central city gentrify by Education and Occupation with 24.7% of its central city gentrifying along those two dimensions.

Conclusion

The previous three chapters each focused on measuring a particular type of gentrification. Chapter 4 focused on measuring income gentrification, Chapter 5 focused on measuring educational gentrification, and Chapter 6 focused on measuring occupational gentrification. This chapter shifted the focus to measuring how common the various combinations of the three types of gentrification are. With three ways for a tract to gentrify, there are seven different gentrification combinations. First, the tract could experience only one type of gentrification (Income-Alone, Education-Alone, or Occupation-Alone). Second, the tract could experience two of the three types of gentrification. There are three ways for a tract to experience two types of gentrification: Income-Education, Income-Occupation, and Education-Occupation. Finally, the tract could experience all three types of gentrification (Complete Gentrification).

This chapter has cataloged how common the various combinations of gentrification were in each of the four decades. The primary results from this chapter are as follows:

- Occupation-Only gentrification was the most common form of gentrification for both slow and rapid gentrification in every decade. However, the percentage of gentrifying tracts that experienced Occupation-Only gentrification decreased in each decade.

- Education-Occupation gentrification was the second most common type of gentrification for both slow and rapid gentrification in every decade. Similarly to Occupation-Only gentrification, the percentage of gentrifying tracts experiencing Occupation-Education gentrification decreased in the 1980s relative to the 1970s and in the 1990s relative to the 1980s. However, unlike Occupation-Only gentrification, there was an increase in the share of gentrifying tracts experiencing Occupation-Education gentrification in the 2000s relative to the 1990s.
- For slowly gentrifying tracts, the least common form of gentrification was Income-Education in every decade. For rapidly gentrifying tracts, Income-Education was the least common form of gentrification in three of the four decades. The exception was the 1980s when Income-Occupation was slightly less common than Income-Education for rapidly gentrifying tracts.
- The largest increase in the percentage of gentrifying tracts experiencing a particular type of gentrification was for Complete gentrification. For slow gentrification, the percentage of gentrifying tracts that completely gentrified increased from 4.9% in the 1970s to 21.0% in the 2000s. For rapid gentrification, the increase was from 1.6% in the 1970s to 13.2% in the 2000s.
- The largest decrease in the percentage of tracts experiencing a particular type of gentrification was for Occupation-Only gentrification. For slowly gentrifying tracts, the percentage experiencing Occupation-Only gentrification decreased from 55.6% in the 1970s to 25.1% in the 2000s. For rapidly gentrifying tracts, the percentage decreased from 67.2% in the 1970s to 39.2% in the 2000s.
- Among slowly gentrifying tracts, Occupation Only was the most common type of gentrification in every Census region in the 1970s and the 1980s. In the 1990s, however, Occupation-Education gentrification was the most common form in the West, while Occupation Only remained the most common form in the other three regions. In the 2000s, Occupation-Education was the most common type of gentrification in both the Northeast and the West, while Occupation Only was the most common type in the Midwest and the South.
- Among rapidly gentrifying tracts, Occupation Only was the most common type of gentrification in every Census region in every decade.

8

CONCLUSION

This book has developed a methodology that could be used to measure the amount of gentrification activity occurring in U.S. cities from 1970 to 2010. The methodology is robust enough to be used at different points in time and can be applied to multiple cities. This makes it possible to provide an analysis of gentrification levels in U.S. cities that is both historical and cross-sectional. Thus, the most important contribution that this book makes to the gentrification literature is to provide a glimpse into such things as how common gentrification has been in U.S. cities, how gentrification levels have changed over time, and which U.S. cities experienced the most gentrification activity over time.

After the introduction provided by Chapter 1, Chapter 2 gives a brief survey of the literature regarding the quantitative analysis of gentrification. The goal was to summarize the various methods that have been used to identify gentrifying neighborhoods with quantitative analysis. A key takeaway from Chapter 2 is that identifying gentrifying neighborhoods is a two-step process. First, it is necessary to identify the neighborhoods that are eligible to gentrify ("gentrifiable" neighborhoods). In this study, gentrifying neighborhoods are central city census tracts with average household incomes that are less than 80% of the median average household income of the census tracts in the metropolitan area.

Once the neighborhoods that have the potential to gentrify have been identified, it is necessary to establish the criteria that will be used to separate the neighborhoods that gentrified from those that did not. Gentrifying neighborhoods are typically identified by the change in a variable or group of variables over time. In this study, income, education, and occupational gentrification are identified individually in Chapters 4–6. In each case, a gentrifiable neighborhood is deemed to have slowly gentrified if its change in the relevant variables exceeds the metropolitan-level

DOI: 10.1201/9781003217459-8

change, while it is identified as rapidly gentrifying if its change in the variable of interest exceeds the metropolitan change by more than 50%.

Chapter 3 provides an overview of the socioeconomics changes that were taking place in U.S. cities and metropolitan areas from 1970 to 2010. The key takeaway from the analysis in Chapter 3 is that the gentrification activity that is identified in Chapters 4–6 occurred during an era where the suburban areas of U.S. metropolitan areas were experiencing growth in population, employment, income, and educational attainment that exceeded those of the central cities. Thus, the gentrification that took place between 1970 and 2010 ran counter to the prevailing trends in U.S. metropolitan areas during that time.

In Chapters 4–6, the methodology that was developed in Chapter 2 is applied to a sample of 100 U.S. cities for every decade from 1970 to 2010. Occupational gentrification is identified as the most common form of gentrification in each decade with income gentrification being the least common form. All three forms of gentrification increased in frequency with each subsequent decade so that gentrification activity peaked in the 2000s. Income and educational gentrification levels increased much more rapidly than occupational gentrification levels. All three types of gentrification experienced substantial increases in the 2000s relative to the 1990s. Denver was identified as the city most significantly impacted by the combination of income, educational, and occupational gentrification from 1970 to 2010.

Chapter 7 uses the results from Chapters 4–6 to identify which types of gentrification were the most common between 1970 and 2010. In every decade, Occupation Only was the most common form of gentrification, while Education-Occupation was the next most common. However, the percentage of gentrifying tracts that experience these two types of gentrification decreased substantially over time. At the same time, Complete gentrification experienced the largest increase in its share of gentrifying tracts over time. Over time, there was a large increase in the number of tracts that gentrified by income, education, and occupation in the same decade.

The analysis in this book has documented that there was a surge in gentrification in U.S. cities during the 2000s. Income, educational, and occupational gentrification levels all peaked in the 2000s. However, one additional insight from the analysis in the book is that the surge in gentrification levels in the 2000s was the culmination of a multidecade trend. All three types of gentrification experienced their lowest levels in the 1970s and all three types experienced a decade-by-decade increase in the number of tracts that gentrified.

When thinking about future directions for quantitative research into gentrification, there are two avenues of research that would complement this study. First, it would be helpful to extend the study to include the decade from 2010 to 2020. At the time of writing, there were serious concerns about the reliability and comparability of 2020 Census data. If these concerns prove to be unfounded or another reliable source of such data is found, then it would be interesting to extend the study to include the data of 2020 and determine the extent to which the trends from the 2000s continued. Second, this study does not address the housing market in any way. Clearly, gentrification also affects housing values and rents and it would be helpful to include a housing dimension in addition to looking at income, education, and occupation. Unfortunately, the Neighborhood Change Database that is used in this study does not provide sufficient housing data for 1970 and 1980 to include an analysis of the housing market in this study. However, if there is reliable data on housing values and rents for the entire timeline, it would be very interesting to apply a methodology similar to the one used in this study to identify neighborhoods experiencing housing gentrification.

INDEX

Printed in the United States
by Baker & Taylor Publisher Services

Printed in the United States
by Baker & Taylor Publisher Services